Bomberg

NEW INSTRUMENTS FOR ENVIRONMENTAL POLICY IN THE EU

New Instruments for Environmental Policy in the EU offers a comprehensive analysis of the debate on new forms of environmental regulation in the European Union. Experts from the fields of law, political science and economics provide a detailed examination of new environmental instruments in six member states – the UK, Germany, the Netherlands, Belgium, Spain and Italy – as well as those adopted at the EU level. The contributors explore the conceptual implications of shifting from a traditional regulatory strategy to one which incorporates new instruments and reveal how key actors, including governments, industry groups and environmental non-governmental organisations (NGOs), view the desirability and feasibility of such a shift.

The conclusions draw attention to critical aspects of instrument design, as well as to the difficulty of accommodating national policy diversity without contravening EU and international trade rules. Drawing on critical research and practical experience, this book presents a number of recommendations for improving the next generation of environmental policies and explores comparisons between the search for new environmental instruments in the EU and similar regulatory transitions worldwide.

Jonathan Golub is a Lecturer in Politics at Reading University, and a former Research Fellow at the European University Institute, Florence, Italy, and the Max Planck Institute, Cologne, Germany.

ROUTLEDGE/EUI ENVIRONMENTAL POLICY SERIES

DEREGULATION IN THE EUROPEAN UNION: ENVIRONMENTAL PERSPECTIVES
Ute Collier

GLOBAL COMPETITION AND EU ENVIRONMENTAL POLICY
Jonathan Golub

NEW INSTRUMENTS FOR ENVIRONMENTAL POLICY IN THE EU
Jonathan Golub

NEW INSTRUMENTS FOR ENVIRONMENTAL POLICY IN THE EU

Edited by Jonathan Golub

London and New York

First published 1998
by Routledge
11 New Fetter Lane, London EC4P 4EE
Simultaneously published in the USA and Canada
by Routledge
29 West 35th Street, New York, NY 10001

Reprinted 2000

Routledge is an imprint of the Taylor & Francis Group

Typeset in Garamond by Routledge
Printed and bound in Great Britain by Biddles Short Run Books, King's Lynn

British Library Cataloguing in Publication Data
A catalogue record for this book is available from the British Library

Library of Congress Cataloguing in Publication Data
A catalogue record has been requested

ISBN 0–415–15696–3

CONTENTS

LIST OF TABLES

NOTES ON CONTRIBUTORS

Susana Aguilar Fernández is a Lecturer at the Complutense University of Madrid. She has published a number of articles dealing with Spanish and German environmental policy, regional government and state institutional design.

Hans Bergman is a national expert in the Economic and Environmental Forward Studies Unit, DG XI, Commission of the European Communities, and formerly worked at the Ministry of Environment and at the Environmental Protection Agency in Sweden. He has published articles on emissions bubbles and on environmental taxation.

Jan Willem Biekart is a policy analyst at the Netherlands Society for Nature and Environment. He has published on a wide range of issues, including strategic environmental management, environmental accountability of industry, environmental agreements and company environmental reporting, and is the author of two books on sediment pollution.

Wolfram Cremer is a scientific Assistant at the University of Rostock. He has published widely on topics of European Community law and constitutional law and is the author of *Forschungssubventionen im Lichte des EG-Vertrages* [Subsidies for scientific research in the context of the EC Treaty] (Nomos, 1995).

Kurt Deketelaere is Professor at the Catholic University of Leuven and Director of its Institute for Environmental and Energy Law. He is the author of several books and articles on international, European, national and regional environmental law, and the editor of *Environmental Law in Belgium: Status Questionis* (Die Keure, 1997; title translated from Dutch).

Jos Delbeke is Head of Unit, Economic Analysis and Environmental Forward Studies, DG XI, Commission of the European Communities, and formerly worked at the IMF. His publications include many articles on environmental economics and related issues.

Eva Eiderström is the Head of Unit for the Shop and Act Green Campaign at the Swedish Society for Nature Conservation. She has published on aspects of ecolabelling in Sweden.

Andreas Fisahn is a scientific Assistant at the University of Bremen, Institute for Environmental Law. He has published on topics of environmental law, constitutional law, and the theory of law and politics. He is author of *Eine Kritische Theorie des Rechts – Zur Diskussion der Staats- und Rechtstheorie von Franz L. Neumann* [A critical theory of law – Franz L. Neumann's theory of law and the state] (Shaker, 1993) and *Niedersächsisches Personalvertretungsgesetz – Kommentar für die Praxis* [The Lower Saxony law of interest representation of public employees – a commentary for practitioners] (Bund-Verlag, 1995).

Jonathan Golub is a Lecturer in Politics at the University of Reading, and former Research Fellow in the Robert Schuman Centre, European University Institute, Florence, and the Max Planck Institute for the Study of Societies, Cologne. He has published widely on topics of environmental policy, EU integration and the interaction of national courts with the European Court of Justice. He is the author of *Hard Bargains: Britain, the Environment and European Integration* (Pinter, forthcoming).

Chris Hewett is a Research Fellow at the Institute for Public Policy Research, London, and former Coordinator of SERA, an environmental non-government organisation affiliated to the Labour Party. He is co-editor of *Energy '98: Competing for Power* (IPPR, 1996).

Duncan Liefferink is a Research Fellow in the Department of Sociology at the University of Wageningen, the Netherlands. He is the author of *Environment and the Nation State: The Netherlands, the European Union and Acid Rain* (Manchester University Press, 1996).

Alberto Majocchi is a Professor of Public Finance at the University of Pavia and former advisor to the Italian government and the European Commission. He has authored and edited numerous publications dealing with fiscal federalism, economic problems of European integration, environmental economics, and environmental aspects of international trade policy.

Karola Taschner is an official of the European Environmental Bureau, Brussels.

Stephen Tindale is a Senior Research Fellow at the Institute for Public Policy Research, London, and former Policy Advisor to the Labour Party. He has written extensively on environmental issues in the UK and is author of *Green Tax Reform* (IPPR, 1996).

PREFACE

Environmental policy in many parts of the world is currently undergoing a fundamental transition, from a traditional command and control approach towards one which places greater reliance on a new arsenal of flexible and efficient policy tools. This transition has been vigorously supported by the European Commission and the Organisation for Economic Cooperation and Development (OECD), but also reflects concern within member states that command and control methods forego potential environmental improvements and impose unnecessary regulatory burdens. Thus the appeal of new instruments, such as market-based approaches and voluntary agreements, which have the potential to reduce the cost of environmental regulation while improving environmental protection.

This book grew out of a summer workshop held on these topics in July 1995 by the Robert Schuman Centre of the European University Institute in Florence, Italy, a postgraduate research institute set up by the member states of the European Union. The workshop and the subsequent completion of the book benefited from the generous financial support of DG XI in the European Commission. The book is highly interdisciplinary, and brings together experts from the fields of law, political science and economics in order to address three central questions. First, what are the conceptual and substantive implications of shifting from a traditional regulatory strategy to one which incorporates new environmental instruments? Second, how do governments, international organisations, industrial groups and environmental non-governmental organisations (NGOs) view the desirability and feasibility of such a shift? Third, what lessons can be learned from the practical experience with new instruments in the European Union in order to improve the next generation of policies?

This book seeks to contribute to the policy debate on several levels. The opening chapter introduces the central issues surrounding the utilisation of new instruments, provides a comparative overview of the subsequent chapters and highlights some of the book's implications. Part I (Chapters 2–7) then offers a detailed analysis of the development and effectiveness of new instruments in six member states. Part II (Chapters 8–11) explores experiences with new instruments adopted at the EU level. Contributors focus particular attention on contentious aspects of instrument design, as well as on the difficulty of accommodating national policy diversity without undermining the EU common market or contravening international trade rules. While the authors do not purport to analyse each and every new tool which has been implemented at national and EU level, their breadth of analysis and interdisciplinary approach should contribute substantially to the growing literature on environmental policy within the EU, while also fuelling comparative analysis with new environmental instruments implemented in other parts of the world.

The book is written for a wide audience of students, researchers and policy-makers. It is intended as a central text for both undergraduate and graduate courses on environmental policy and politics. It is also designed to be incorporated into courses on comparative public policy, West European politics, and the European Union. Policy-makers and researchers will find the book useful both for its detail and comprehensiveness, and as a companion to the other volumes in the EUI Environmental Policy Series, *Deregulation in the European Union: Environmental Perspectives* (U. Collier, ed.) and *Global Competition and EU Environmental Policy* (J. Golub, ed.).

Jonathan Golub
Florence, August 1997

1

NEW INSTRUMENTS FOR ENVIRONMENTAL POLICY IN THE EU

Introduction and overview

Jonathan Golub

European environmental policy is currently undergoing a major transition. At the supranational level, both the Maastricht treaty and the Fifth Environmental Action Programme herald a new era dominated by the search for more flexible and efficient instruments to replace traditional forms of regulation. At the national level, this search has been under way for some time in many of the respective member states of the European Union (EU).

This chapter introduces the main issues surrounding the use of new instruments, and offers an overview of their successes and failures. The first section briefly reviews the dominance and shortcomings of the traditional command and control approach in European environmental policy. The second section has two parts: the first identifies the various types of new instrument considered throughout this volume and discusses their reputed advantages over command and control; the second identifies the domestic and international forces which have been instrumental in getting new policy tools put into practice. The third section opens with a discussion of various criteria for gauging the effectiveness of new instruments. A comparative analysis of member state experience then highlights factors which have facilitated successful application of each type of instrument, as well as those which have undermined their effectiveness. To gain a better perspective on developments within the EU, this section is followed by a brief examination of how new instruments have been used in the United States and Japan. The final section draws out some of the broader implications this volume's findings hold for the future of environmental policy design, the nature of governance within the EU and the prospects for maintaining global free trade.

Command and control

Originating in the late 1960s, the first generation of environmental policies throughout the member states of the EC primarily followed the so-called 'command and control' approach. The command and control approach is characterised by direct regulation: the government prescribes uniform environmental standards across large regions, mandates the abatement methods required to meet such standards, licenses production sites which adopt the required methods, and assures compliance through monitoring and sanctions.

The chapters in this volume make clear that the prevalence of command and control instruments built up throughout the 1970s generally reflected a north–south 'green' divide (see also Collier and Golub 1997). Germany and the Netherlands boast particularly strong environmental records, and had extensive direct regulation covering air, water, waste and noise in place by the mid-1970s. Over the years each of these states set and tightened multitudes of emission standards based on what could be achieved by using the 'best available technology' (BAT).

The first generation of British environmental policy incorporated greater discretion and flexibility, but the corpus of legislation built up since the nineteenth century was grounded firmly in the command and control tradition. Although less demanding than BAT, British regulation to control air pollution set emission limits according to the 'best practicable means' and periodically tightened these limits in line with technological developments (McCormick 1991; Golub 1996a). For water, direct regulation made authorities responsible for ensuring the 'wholesomeness' of rivers, frequently taken to be World Health Organisation (WHO) standards. Industrial discharge consents were then generally structured so that overall pollution levels would not exceed WHO recommendations. Command and control also pervaded Belgian environmental policy prior to the early 1980s. While these laws did not require the use of BAT, a variety of direct regulations set national and regional quality objectives in each sector (air, water, waste, etc.).

While the four chapters exploring environmental instruments in northern Europe highlight variations within a group of states sharing strong green traditions and extensive bodies of command and control regulation, the chapters on Spain and Italy (Chapters 6 and 7) reaffirm the north–south dichotomy, illustrating that environmental protection has always occupied a less auspicious position in the highly fragmented policy-making of Mediterranean states (see also Collier and Golub 1997; La Spina and Sciortino 1993). Nevertheless, while regulation devised in the 1960s and 1970s was extremely patchy in southern Europe, amounting to no more than a handful of air and water laws, it too reflected a command and control approach.

In many cases the direct regulation found within European states arose from external constraints, as the expanding corpus of environmental policy adopted at the EC level since 1973 also relied heavily on command and control. EC regulations included many provisions requiring existing industrial installations to curtail pollution in accordance with uniform emissions standards and requiring all new plants to apply BAT.[1]

Particularly in Spain and Italy, first generation national command and control instruments were put in place to satisfy the requirements of EC law, filling a pre-existing regulatory void. But even in the north, EC environmental legislation sometimes reshaped or solidified previous national measures. Prescribing uniform emissions standards and abatement technology throughout a state or throughout the EC, with attendant economic costs, was considered by many an equitable response to problems of upstream–downstream pollution, preferable to simply setting quality targets which privileged those states, regions or firms fortunate enough to avoid cross-border effluent problems. Closely bound up with the emphasis on uniformity and technical solutions was the perception that allowing industry in peripheral, isolated or sparsely populated regions to meet environmental objectives at relatively less cost than those in other parts of the EC would constitute an unfair competitive advantage which would distort the common market.

Two decades of experience revealed a number of regulatory failures associated with the traditional command and control approach. These shortcomings fall into three categories: economic inefficiency, environmental ineffectiveness and democratic illegitimacy (see Tietenberg 1988; Eckersley 1995a).

By its very nature, command and control tends towards economic inefficiency by imposing uniform reduction targets and technologies which ignore the variable pollution abatement costs facing individual firms. In practice, marginal costs of pollution reduction vary widely among industries, depending on factors such as age and location of plant. From an efficiency perspective, the result is that some firms regulate too much, others not enough. Installing state-of-the-art equipment or cutting emissions by a specified amount (for example by three tonnes of sulphur per month) will necessarily be easier and cheaper for some plants than for others. While a designated overall level of pollution reduction could thus be achieved at less overall cost if abatement efforts were concentrated where marginal costs were lowest, command and control discourages such efficiency.

In some cases the command and control approach also introduces inefficiency by eschewing environmental quality objectives in favour of uniform emission standards. Consider the case of two identical firms, one located in a sparsely populated area, the other in a city. Because of the differing capacities of their local ecosystems to absorb additional units of pollution, the firms face variable abatement costs and would therefore have to make drastically different reductions in their emissions to meet the same quality targets. Uniform emissions standards and BAT produce inefficiency by imposing identical costs. Not allowed to take advantage of its favourable location, the rural firm is forced to overregulate and jeopardise its economic performance; at the same time, even tougher standards and penalties beyond BAT might actually be required for the urban firm to meet acceptable pollution levels or compensate for environmental damage.

Finally, the command and control approach contributes to inefficiency because it stifles incentives to reduce emissions beyond mandated levels and to develop innovative pollution control technology. Rather, the prescription, license and monitor approach generates a static situation where, having installed a designated technology or achieved a certain level of emissions, polluters would only incur unilateral costs and competitive disadvantages from further reductions. BAT rules are particularly effective in stifling innovation, because firms are forced to adopt expensive equipment regardless of whether other, and sometimes more radical, solutions might be found at less cost (von Weizacker 1990: 202).

Command and control is not only an expensive approach to pollution reduction, but one which, according to many analysts, has also reached the limits of its environmental effectiveness. To a large extent these are related issues, as adoption of new and increasingly stringent emissions limits and technologies intended to safeguard the environment encounter intense political resistance when they are seen as excessively costly. But regulatory failures also plague current policies, for instance when fiscal austerity forces governments to curtail the expensive oversight and enforcement mechanisms vital to command and control, resulting in rampant non-compliance.

A third criticism levelled at the command and control approach is that it lacks democratic legitimacy (Eckersley 1995b; Dryzek 1995). Regulatory bodies responsible to the public for identifying environmental problems, standard setting, determining what constitutes the best available technology and enforcing compliance develop close and often dependent relationships with industry because of the latter's detailed knowledge of, and direct interest in, polluting activities and potential abatement options. As currently constructed, command and control instruments fail to alleviate the information asymmetries which effectively exclude the general public and environmental interest groups from the decision making process and allow polluters to 'capture' regulators, thereby shaping or blocking environmental policies in accordance with their own economic self-interest.[2]

The shift towards new instruments

An arsenal of new instruments

The limitations and regulatory failures of the traditional command and control approach have sparked a search at both the national and the EU level for a second generation of instruments which promises greater flexibility, efficiency and effectiveness.[3] States have experimented with an impressive range of new tools, some of which might be classified as 'economic' and 'suasive' instruments (OECD 1994b: 16). The former category includes several types of environmental taxes and charges, comprehensive 'green tax reform', tradeable pollution permit systems, government subsidies for environmental improvement and deposit/refund schemes. The latter group consists of ecolabels, ecoaudits and voluntary environmental agreements. A third class of new instruments which receives less attention in this volume focuses primarily on altering liability and insurance rules in a manner which benefits the environment. An extensive literature discusses the economic theory underpinning this wide range of potential new tools, as well as their individual merits (Tietenberg 1988; Helm and Pearce 1990; Hahn 1993; Teubner *et al.* 1994; House of Lords 1993).[4]

New instruments are intended to provide the efficiency and positive incentives which command and control lacks. Taxes and charges levied on each unit of emission force firms to internalise the costs of their pollution, thereby better achieving the 'polluter pays' principle, but they also allow industry the freedom to optimise its reduction methods. Tradeable permits and voluntary agreements achieve the same end by setting long-term environmental goals without prescribing specific abatement technology. Latitude to find the most cost-effective means of reducing pollution improves static efficiency but, more importantly, it yields dynamic efficiency by providing incentives for firms to pursue constant pollution reduction and technological innovation (Carraro and Siniscalco 1993; Eckersley 1995b: 9; EC 1992, 1993, 1994). Each successive reduction in emissions saves the firm from paying tax, allows it to sell its tradeable permits to 'dirtier' competitors or allows it to meet future targets at lower cost.

Compared to command and control, ecolabels and ecoaudits also substantially reduce the regulatory burden on firms because they prescribe neither targets nor technologies. Rather, these instruments generate incentives for pollution reduction by harnessing the market power of 'green consumerism'. Armed with the information which labels and audits provide about the content and manufacturing process of products, and about the internal environmental performance of firms, consumers will be able to express their preference for environmentally friendly behaviour, and firms will be forced to respond accordingly or else lose market share.

Ecoaudits also provide several other potential advantages over command and control. First, industry expects to improve relations with green consumers, banks and insurance companies by publicising official certification of its environmental commitment. Environmental audits can also reveal new knowledge about a firm's production process, and they often illuminate means of improving efficiency and curtailing consumption of natural resources, both of which result in substantial savings. Compared with traditional forms of direct regulation and BAT rules, ecoaudits can also introduce a desirable dynamic element, providing even the worst polluters with positive incentives for future environmental improvement.

Proponents of new instruments suggest that governments also stand to profit by reducing their reliance on command and control, because market-based and suasive tools require less expenditure on implementation and oversight. Furthermore, green taxes might represent a politically popular and lucrative source of government revenue.

Another important advantage attributed to new instruments is that they decrease regulatory capture and lend legitimacy to environmental policy by substituting direct public involvement for command and control's infamous 'poacher and gamekeeper' relationship between industry and regulatory bodies. Labels, audits and ecotaxes allow consumers to decide for themselves the value they place on environmental improvement. Voluntary agreements also have the potential to improve the legitimacy of environmental policy by involving industry in close consultation rather than open confrontation with government. Widening the range of parties participating in the consultation process to include green groups and local community officials further enhances the legitimacy of eventual decisions. Japanese experience with voluntary agreements suggests that industry is willing to disclose important information necessary to set ambitious targets because it considers them economically feasible, and also because it benefits from the trust developed with local communities (Rehbinder 1994).

The widespread appeal of new instruments

The use of new instruments varies widely among member states, as do the reasons for their introduction. The chapters in this volume illustrate the range of domestic and international pressures animating the shift away from command and control: the ascendance of new economic paradigms and political ideologies, demands from industrial groups, the agendas of certain political parties, pressure from environmental or academic organisations, and the requirements of EC law. While in most cases all three of the above considerations (cost, effectiveness and legitimacy) influenced policy change, each author has attempted to identify more precisely the timing of reform and the interplay of domestic and international politics which yielded new tools. For example, academics and think-tanks have played a particularly important role in Britain and Belgium, while EC law has been the driving force in Spain and Italy.

Industry groups have been instrumental in placing new policy tools on the political agenda. Their primary concern, not surprisingly, is that the ability of EC firms to compete in the global economy requires a lightening of the 'burden of regulation' (EC 1995, part 2: 10), which poses unnecessary 'costs, rigidities and obstacles to innovation' (*ibid.*: ii). Industry supports most alternatives to direct regulation, arguing that they will improve environmental protection while reducing compliance costs, but has expressed a strong preference for suasive instruments (information schemes and voluntary programmes) over other types of new tools (*ibid.*, part 2: 2, 52; UNICE 1993; but see Hahn 1995: 151–2).

However, industry has also exhibited some ambivalence towards the shift, in part through sheer hypocrisy, but also because of the uncertainty and the transaction costs they face by abandoning what is often a well understood regulatory approach over which they exercise considerable influence. In some cases (discussed below) the industry perspective on new instruments involves mixed motives, for instance when dealing with free-rider problems or when the interests of large and small firms diverge.[5]

An interesting theme which recurs throughout the country studies is the gradual conversion of environmental non-governmental organisations (NGOs) and Green Parties from an initial position of scepticism regarding new instruments towards one of guarded enthusiasm. In Germany, the Green Party was split in the late 1970s and early 1980s between pragmatic and fundamentalist factions, only the former of which supported gradual reform and experimentation with new tools. Over time, this group and its agenda gained control of the Green Party as fundamentalist objectors were marginalised (see Chapter 3). Italian NGOs resisted new tools in the 1970s, fearing that they would provide insufficient environmental protection and objecting on ethical grounds to what they viewed as a 'right to pollute', but have since given full support to new instruments as a means of overcoming the regulatory failures of command and control (see Chapter 7). UK NGOs have consistently opposed new instruments, but seem recently to have accepted them as inevitable and turned their attention towards ways in which the regressive effects of green taxes might be alleviated (see Chapter 2).

But enthusiasm has its limits, and a number of NGOs have cautioned that a rush towards new instruments should not lose sight of the fact that re-regulation, rather than indiscriminate deregulation, is required (EEB 1995: 4). The highly publicised Molitor report on 'legislative simplification' in the EU (EC 1995) received a particularly cool reception from environmental NGOs, who feared that it heralded a reckless repatriation of environmental competences at the expense of necessary supranational legislation (EEB 1995). It should be noted, however, that the report's recommendations even elicited a dissenting opinion within the 'group of experts' which admonished the Commission to replace direct regulation with new instruments on a selective basis in order to safeguard the environment, rather than to fulfil an ideological deregulatory objective.[6]

International pressures have also played an important part. Since the early 1970s the OECD has consistently promoted widespread use of new environmental instruments (Eckersley 1995b: 9–10; OECD 1975, 1989). The supranational institutions of the EC, particularly the Commission and the

European Parliament, have expressed similar overwhelming support, and new tools figure prominently in the Community's most recent (fifth) Environmental Action Programme (EP 1994: 12–14; EC 1992: 6–7; 1993, 1994). Commission documents cite many advantages of new instruments over the traditional command and control approach, including their greater flexibility, their cost-effectiveness and the incentives they provide for firms to internalise the negative externalities associated with pollution. Moreover, the shift away from command and control provides a greater sensitivity for the varying absorption capacity of regional ecosystems, the exploitation of which, according to the Commission, constitutes an 'entirely legitimate source of comparative advantage' (EC 1992: 5; EC 1996b: 7, 10).

Despite its enthusiastic support, however, the EU has been less successful in adopting new tools at the supranational level than it has been in encouraging national experimentation with a variety of market-based and suasive instruments. The effectiveness of these instruments and the problem of accommodating national diversity within a common EU framework are taken up in the next section.

Gauging the effectiveness of new instruments

There is widespread agreement that the effectiveness of new instruments depends largely on their ability to harness market forces in favour of environmental protection. There is considerably less agreement, however, on what constitutes an acceptable level of pollution and the extent to which market forces should be entrusted to determine this for society. How one answers these questions fundamentally alters the optimal design of any instrument and in part determines the appropriate standards for judging its success.

The so-called 'free-market environmentalism' (FME) debate (Eckersley 1993) exemplifies the competing philosophies among proponents of new instruments. Supporters of FME favour reducing pollution to an 'optimal' level – defined as the amount of reduction which consumers are willing to pay for once the full social costs of negative externalities are reflected in market prices – while their opponents advocate pollution levels consistent with sustainable growth. The latter levels are far lower than a market equilibrium would render and therefore depend on purposeful government imposition of targets. The radical FME perspective is often, but not exclusively, espoused by conservative or libertarian American think-tanks (T. Anderson and Leal 1991; F. Smith and Jeffreys 1993).

The extreme FME position enjoys little support throughout the EU. Rather, a broad consensus of actors advocates re-regulation rather than deregulation because a totally 'free' market misallocates natural resources and produces inadequate incentives to prevent environmental degradation.[7] The major players accept the proposition, which Weale suggests 'is so banal as to require little justification' (Weale 1993: 158), that 'the philosophy underlying [new] instruments is that the market should be used for economic efficiency purposes in a merely instrumental way in order to achieve environmental policy goals that continue to be set by the state' (Rehbinder 1994: 147). Even the Molitor report advocates that new tools should 'be designed to achieve a required level of quality', expressed as targets set by government (EC 1995: 14).

This suggests that the design and effectiveness of new environmental instruments should be judged on three criteria: first, how well they curtail pollution to levels consistent with sustainable development; second, the savings they render in compliance costs; third, the extent to which they enhance the legitimacy of environmental policy making. These criteria mirror the shortcomings of conventional command and control regulation. As the chapters make clear, however, these criteria are largely independent and frequently conflictual. Moreover, their order of importance is ranked differently by various interested parties.

Green taxes and subsidies

Environmental taxes on polluting activities can take many forms, and are designed to raise revenue, alter the behaviour of polluters, or both. The most widespread use of charges in the EC has been simply to raise cash or to cover the operating costs of treatment plants. Revenue-raising taxes have been used by most states, in both the north and south, to finance treatment plants and cover the administrative costs of controlling water, air and waste pollution.

In contrast to revenue raising taxes, incentive taxes seek to encourage environmentally friendly production and consumption patterns. Differential tax rates (for example the 'fuel escalator' in the UK) have been widely used, particularly to encourage greater use of unleaded petrol and low-sulphur fuel oil. Tax differentials have also been established, mostly in the northern member states, to subsidise the purchase of cars which meet stringent emissions standards several years before these standards take effect legally. Five member states have already adopted carbon-dioxide (CO_2) taxes, while others, including Italy, are making theirs conditional on adoption of a similar instrument at the EU level. British and Italian charges on waste deposited in landfill sites constitute an interesting new example of green incentive taxation, and many of the chapters identify a range of other proposed environmental incentive taxes, for instance vehicle taxes and motorway tolls in the UK and a sustainable tourism tax in Italy.

One of the broadest and most ambitious attempts at green incentive taxation was made by Belgium in 1993. As Deketelaere explains, the 'ecotax' was intended to alter consumer behaviour substantially, over a wide range of products, but industry resistance and poor initial design undermined its implementation and have partly limited its effectiveness (see Chapter 5). Nevertheless, re-use of packaging, batteries and cameras has increased, suggesting that industry responded to provisions in the tax which offered them exemptions for establishing recycling systems. As in Belgium, taxes have been used to support recycling not only in many of the northern member states, but also in Spain and Italy, partly as a means of implementing the EC Directive on Packaging Waste.

When taken together, do all of these individual taxes amount to a substantial 'greening' of national tax systems? Unfortunately not. In 1993 environmental taxes contributed only 1.5 per cent of the total taxes raised in the EU, while energy taxes comprised 5.3 per cent, and only the latter has risen noticeably since 1980 (EEA 1996: 24–6). In fact, for Italy and the UK, the proportion of total tax revenue provided by environmental and natural resource taxes was significantly higher in 1970 than in 1990 (*ibid.*: 51). Moreover, the modesty of these figures does not reveal the full extent of the problem. Existing revenue raising taxes do not always fully cover the operating costs of treatment plants and, even when they do, cost recovery by itself does not indicate that sufficient pollution reduction is taking place. Britain, for example, does not tax the actual volume of pollution treated. In Spain, water taxes have generally failed – many firms and municipalities have never paid, and many regions have simply not implemented the tax provisions (see Chapter 6). Also, incentive taxes, even when implemented, fall far short of inducing producer and consumer behaviour consistent with sustainable development. Clearly, a much more aggressive and comprehensive system is required.

Although many states are considering comprehensive environmental tax reform, and five have established special tax commissions for this purpose (Belgium, Denmark, Finland, the Netherlands and Sweden), it remains an elusive goal. For one thing, states have not dismantled the myriad subsidies and levies which favour environmentally harmful activities and externalisation of pollution costs, such as the use of dirty coal, airline fuel and pesticides (see Chapter 11; EC 1996b, 1996c). These cases illustrate the importance of dealing with environmentally harmful activities which currently go untaxed.

Widespread use of green taxes has also been plagued by two important distributional problems: the fear of eroding economic competitiveness and the regressive nature of certain taxes (EEA 1996: 33–4; OECD 1994a; S. Smith 1995). To deal with regressiveness, many national tax schemes incorporate cut-off points and exemptions so that the poorer segments of society are not deprived of essential goods such as water and electricity. Similarly, Italian energy taxes vary substantially according to region, thereby offsetting regressive impacts in the south (see Chapter 7). Failing to incorporate such mechanisms can have a high political price. A British proposal in 1993 to raise the level of value-added tax (VAT) on fuel from 8 per cent to 17.5 per cent was defeated mainly because it would have imposed disproportionate costs on lower-income households. In 1997 the new Labour government slashed the rate back to 5 per cent.

Issues of economic competitiveness have always played a central role in the formation of environmental policy at both the national and the EC level (Golub, 1998b), and now look set to occupy an equally pivotal position within the new generation of environmental instruments, particularly green taxes.[8] As mentioned previously, while industrial groups have been vocal advocates of a shift away from command and control for economic reasons, they have frequently pleaded that green taxes would merely increase their production costs relative to competitors, resulting in lost market share, higher unemployment, depressed growth and stifled investment.

The inability to resolve distribution and competitiveness issues among twelve (now fifteen) member states, combined with unanimous voting requirements and considerations of subsidiarity, has prevented the adoption of green taxes at the EU level, the deadlocked carbon-tax proposal serving as the primary example (see Chapter 11; see also Heller, 1998; Golub 1996b). Unable to muster sufficient political support for supranational legislation, the Commission has explored ways of allowing member states greater latitude to implement their own new instruments (see below).

Advocates of ambitious green laws deny this presumed negative relationship between environmental standards and economic competitiveness, and suggest instead that stringent environmental laws, when properly designed, actually promote the competitive advantage of firms – the so-called 'win–win' thesis (Golub, 1998a). This claim is also a central element of the 'ecological modernisation' paradigm (Weale 1992). Two of the fundamental arguments offered in support of the win–win hypothesis have been, first, that because new instruments exhibit proper design they are more cost-effective than traditional regulatory tools, and therefore encourage more efficient use of resources and lower production costs; second, that their design also generates incentives for investment in environmental research and development, the fruits of which can then be sold to competitors, enabling 'first movers' within the EC to capture the lucrative global market for pollution abatement technology and services (EC 1992). The Commission has gone so far as to suggest that 'a stronger reliance on market based instruments is the key' to the success of sustainable development and of efforts to construct a positive relationship between economic growth and the environment (EC 1994: 2). A win–win strategy has also underpinned green tax proposals in several member states, including Italy and the UK (see Chapters 2 and 7).

Even among advocates of new environmental instruments, however, opinion is divided over how to design green taxes which will enhance rather than undermine economic competitiveness. The most pervasive line of thinking suggests that in order to resolve distribution problems, offset costs and spur technological innovation and diffusion it is essential that the proceeds of such taxes be earmarked for specific uses, rather than increasing the overall level of taxation (EC 1996b, 1996c; OECD 1994a; M. Anderson 1994; EEA 1996: 34–6). Revenue can be funnelled directly back to firms in the form of research and development subsidies, or can be used to maintain 'fiscal neutrality', whereby the government retains its overall revenue level while cutting the rates of other taxes payable by firms, such as those on labour and profit. Proponents of this view argue that CO_2 taxes, for example, fail without such research and development (R&D) subsidies (Carraro and Siniscalco 1993).

Experience from the Netherlands illustrates the controversy over alternative uses of green tax revenue. Despite flowing into the general budget, Dutch water and fuel levies have proved effective in reducing pollution (see Chapter 4). But Liefferink notes that this success might be a product of exceptional circumstances and that greater environmental improvement might have resulted from targeted use of the tax revenue. It is thus not surprising that earmarking in these fields is now under consideration. Interestingly, the Dutch CO_2tax, which lacks earmarking, has yielded only a 1 per cent reduction in emissions (EEA 1996). Moreover, Dutch industry has demanded that all new environmental charges should be returned to them in one form or another.

Another way to avoid eroding economic competitiveness is simply to exempt industry from paying green taxes. This became a centrepiece of the EU carbon-tax proposal, and existing or planned carbon taxes in most member states (for instance the Netherlands, Italy and Germany) also include provisions exempting energy-intensive sectors, which would have been hardest hit (EEA 1996). Such exemptions diminish political opposition from the business community, but exact a corresponding environmental price as the most polluting industries escape taxation and thus avoid internalising the full social costs of their actions.

Closely related to recycled tax revenue, another new instrument for improving environmental protection is for governments to provide firms with subsidies in the form of grants and low-interest loans, which are used to purchase clean technology or acquire environmental management expertise. While an earlier OECD report found green subsidies generally negligible (OECD 1990), other studies have concluded that, outside the EU cohesion states, subsidies are substantial and have risen since the early 1980s (Clement 1995). The chapters in this volume suggest a mixed picture. In Belgium subsidies are present but negligible, while in Spain the government has channelled funding for R&D into clean technology. In the UK, which generally eschews subsidies, a fossil fuel levy encourages the use of 'clean' energy sources. However, it has received mixed reviews because 90 per cent of the funds have gone towards nuclear energy production.

Green taxes linked with subsidies might offer an attractive form of new instrument, but there are serious obstacles to this approach. Most importantly, the levying of national environmental taxes, the introduction of differentiated taxes and the earmarked use of tax revenue each have the potential to violate EU, General Agreement on Tariffs and Trade (GATT) and World Trade Organisation (WTO) laws designed to maintain the common market and free international trade (see Chapter 11; Golub, 1998b; Vogel, 1998; Esty 1994). Depending on their design, green taxes can function as discriminatory barriers against foreign firms and products, and the recycling of tax revenue can contravene rules on state aid and competition. This becomes particularly worrying if, as many have suggested, the success of green taxes and of win–win scenarios depends on completely offsetting the cost of environmental charges with earmarked revenue and subsidies. In light of the potential conflict between environmental and common-market objectives, the Commission has undertaken the difficult task of devising guidelines which will serve as a legal framework in which member states might experiment with green taxes without violating EU law (see Chapter 11; see also Grabitz and Zacker 1989). Whether such rules survive WTO scrutiny remains to be seen.

Compared to command and control measures, do green taxes deliver a reduction in administrative costs, as advocates suggest? As with their contribution to environmental improvement, experience reveals less than spectacular results. The margin for cost savings is reduced substantially when one recognises that setting appropriate objectives and charges for new instruments in a deregulated climate demands the same level of information about firms, consumers and environmental degradation as devising BAT and other command and control standards (Heyvaert 1997; Weale 1993). Furthermore, green taxes require oversight, monitoring and enforcement mechanisms appropriate for dealing simultaneously with the environmental behaviour of industries, consumers and households. Sometimes sufficient mechanisms are already in place from previous command and control laws; but not always, and environmental tax regimes entail substantial expenditure when new administrative structures must be designed or old ones reconfigured. Under certain circumstances, oversight of BAT installation might actually prove an easier and more cost-effective approach (Jacobs 1995: 58; EEA 1996: 39). To take two examples from the country studies in this volume, the Belgian ecotax cost more to construct and oversee than it has yielded in revenue (see Chapter 5); and until the full introduction of water metering in the UK, regulators will clearly possess insufficient information to levy and enforce incentive taxes (see Chapter 2).

Voluntary agreements

New instruments can also take the form of 'voluntary' agreements, whereby governments enter into negotiations with industry over the extent and timing of feasible environmental improvement without mandating any particular pollution abatement method. Also known as negotiated agreements or covenants, these instruments have been used in most member states – to meet EC packaging waste goals, to reduce CO_2 and sulphurdioxide (SO_2) emissions, to phase-out chlorofluorocarbons (CFCs) and to improve energy efficiency – but are particularly prevalent in the Netherlands, with its strong tradition of consensual politics (see Chapter 4).

Proponents contend that, compared to command and control, voluntary agreements provide flexibility, cost savings and a sense of regulatory legitimacy, in exchange for which firms will agree to more ambitious environmental goals. But reaping these benefits depends on proper instrument design. Studies suggest that an absence of essential provisions has rendered 50 per cent of Dutch covenants ineffective, while many agreements in the UK suffer from these same deficiencies (see Chapters 2 and 4). Spain has also made use of voluntary agreements, but it is too early to judge their economic or environmental results (see Chapter 6).

Successful voluntary agreements – and Biekart discusses quite a few – share four design characteristics (see Chapter 8). First, they must contain substantive commitments – quantifiable environmental targets and timetables rather than ambiguous industrial promises eventually to reduce pollution. Second, there must be a 'stick behind the door' – the threat of direct regulation if industry fails to meet the covenant's environmental objectives. Third, both the negotiating process which leads to an agreement and its subsequent implementation must be transparent in order to guarantee enforcement as well as legitimacy among the maximum number of concerned parties. Fourth, voluntary agreements should be legally binding.

In many cases, however, agreements do not include these characteristics. Substantial information deficits have hindered the effectiveness of many agreements, and there is little evidence that the situation is set to improve – less than 1 per cent of firms publish annual environmental reports, and industry opposes the creation of pollution registers (see Chapter 8; see also EC 1995: 17). A major problem with previous agreements, including several in the Netherlands and the UK, has been the exclusion of NGOs from all negotiations or their restriction to early stages of discussion (see Chapters 2 and 4; EC 1996a: 28). As Deketelaere points out, without full access by all interested parties, industry domination (and withholding) of information biases the 'agreement' against environmental protection (see Chapter 5).

What accounts for the large number of agreements lacking one or more of Biekart's essential characteristics? A number of political and legal considerations have played a part. The disappointing results from previous 'gentlemen's' agreements did encourage many EC states (including Germany, the Netherlands and Belgium) to shift towards a more legal approach (see Chapters 3–5; EC 1996a: 23–4, 28). In the UK a similar evolution has occurred in the area of pesticide control (Baggott 1986: 64). Despite the environmental attractiveness of greater legal formality, however, such agreements have dubious political viability because industrial groups view them from a position of mixed motives: firms welcome legally binding agreements because they prevent free riding, but nevertheless remain sceptical, fearing that legality sacrifices industry discretion over the timing and means of achieving implementation. Industry's sensitivity to maintaining discretion might prove insurmountable, as in Flanders, where not a single agreement has been concluded since agreements became legally binding in 1994 (see Chapter 5).

The legal nature of an agreement affects government actions as well as those of firms. For industrial groups the attractiveness of voluntary agreements also depends on the government's ability to forego future regulations for a set period of time, a promise frequently made but a legal power which remains very much contested (EC 1996a: 26; Rehbinder 1994). EU legislation injects additional uncertainty, in that BAT requirements, new standards and other disruptions to investment schedules can enter 'through the back door' despite government assurances, undermining industry's reasons for concluding an agreement in the first place. Belgian agreements, for instance, specifically allow for new rules imposed by Brussels (see Chapter 5).

While the lack of targets, transparency, potential regulation and legality explains the failure of many agreements, several other problems have also emerged. One is the difficulty of gauging 'baselines' – the amount of pollution reduction firms would achieve by themselves without a voluntary agreement.[9] Underestimating the baseline will lead to relatively lax targets compatible with status-quo trends and 'business as usual'. This offers minimal environmental improvement, certainly much less than technology would allow.

Another problem concerns the scope for technological innovation and cost savings provided by voluntary agreements. Have agreements allowed industry greater flexibility over the means of reducing pollution, lowered their compliance costs and spurred development of clean technology? The answer is difficult to determine, particularly when some covenants retain provisions requiring firms to adopt a certain type of technology; a sure recipe for stifled innovation according to some (Carraro and Galeotti 1995). On the positive side, Dutch chemical covenants retain BATNEEC but provide greater flexibility on the timing of its implementation. Several companies estimate that they have saved 10 per cent in administrative costs.

As yet there have been no voluntary agreements concluded at the EU level, although the Commission claims that they could be used in many areas. Instead, much as is the case with green taxes, the Commission has developed an EU framework which encourages greater use of voluntary agreements within the member states and stresses the need for proper instrument design along the lines discussed above (EC 1996a).

Ecolabels

Instead of taxes or direct regulation, many states have employed ecolabels, suasive tools designed to achieve similar pollution reduction indirectly by providing consumers with greater information about the environmental qualities of specific products. Polls suggest that many people prefer to buy green products when possible and are even willing to pay a higher price for them; armed with proper information these green consumers can reward environmentally friendlier brands with greater market share (van Goethem 1992). This becomes a perplexing task amidst the current deluge of advertisements extolling the environmental friendliness of nearly every available item. Ecolabelling schemes therefore involve harmonised or standardised procedures and logotypes, often with third-party evaluation, all of which allows consumers to distinguish between 'greenwash' and legitimate environmental claims.[10]

Within the EC, national ecolabels have proliferated since the late 1970s, and consumers were eventually confronted in shops with the German Blue Angel (1978), the Scandinavian White Swan (1989) and the French *NF-Environnement* label (1992), to name just a few. While these national labels achieved a certain level of success, encouraging greener consumption patterns in several member states (particularly in Sweden, as Eiderström notes in Chapter 9), they also generated some environmental and economic problems in the context of extensive intra-EC trade. Not only did it become increasingly difficult for shoppers to discern the merits of competing official labels, so that they would trust a local label regardless of the standards it represented, but the fact that national labelling criteria could be used to discriminate against other EC producers threatened to undermine the common market. States can easily adjust their label criteria to favour domestic products and production processes. And, even without discrimination, EC producers of green products wanting to penetrate neighbouring markets faced enormous transaction costs from having to make separate applications for each national ecolabel.

Many argued that a single European-wide ecolabel would resolve these problems, and it was against this background that the 'EU Flower' was adopted in 1992. However, as Eiderström discusses in Chapter 9, the scheme has been plagued by considerable disagreement within the Commission, as well as among member states, over the development of criteria for individual products. The Commission has undertaken reforms to streamline the process, but as of mid-1997 only five standards had been adopted.

Experience with the EU Flower and various national programmes highlights the fact that, like green taxes and voluntary agreements, the effectiveness of ecolabels hinges on resolving contentious matters of instrument design. Most importantly, there must be consensus on the criteria for conferring an ecolabel: will it reward the reduction of pollution caused during the production process, the omission or inclusion of certain product ingredients, or the curtailing of environmental damage caused during the product's use and disposal? The so-called 'cradle to grave' approach, which applies life-cycle assessment (LCA), represents an attempt to synthesise, rather then arbitrarily weight, these factors. Nevertheless, intense debate has arisen among member states and with the EU's trading partners over how to apply LCA properly.

As in voluntary agreements, openness and transparency is also critical in the design of ecolabel programmes. When all interested parties enjoy access to the criteria-setting process ecolabels can improve the legitimacy of environmental policy by substituting direct consumer and NGO involvement for the agency capture found so frequently under command and control. According to Eiderström, many national ecolabels, as well as the EU Flower, fail to deliver adequate transparency because criteria are devised within standardisation bodies or under other conditions where access to information is dominated by industry. Guaranteeing sufficient access to a wide range of firms, both large and small, domestic and foreign, is also an important factor in building legitimacy and averting trade distortions.

As with many of the new instruments discussed in this volume, ecolabels encounter serious legal impediments with EU and WTO rules. A diversity of national labels, each of which adopts a different form of LCA and neglects access and appeals procedures for firms in neighbouring EU states, will almost certainly generate trade barriers and violations of the EEC treaty's rules on competition. In lieu of convergence among national instrument-design and criteria procedures, the imposition of the EU Flower as a common ecolabel might avert some of these EU legal problems. In an international context governed by WTO rules, however, convergence (or the use of a single ecolabel) might have to include non-EU states, since the EU Flower itself has been attacked by the US, Canada and Brazil as discriminatory and based on improper LCA (Vogel, 1998).

Even if it were possible to surmount the legal and economic obstacles discussed above, it remains unclear whether the optimal environmental solution consists of a single EU label or continued national diversity. Having already made substantial commitments to national labelling schemes, will consumers and industry see the value of an EU label? Even if this means suspending well-known schemes such as the Blue Angel and the White Swan while protracted debate continues over the EU Flower?[11]

Ecoaudits

Environmental management systems (EMASs), often referred to as ecoaudits, are a second type of entirely voluntary suasive instrument. They work as follows: in exchange for official government confirmation of their efforts, firms undertake a comprehensive assessment of their production processes and commit themselves to achieving steady improvement in environmental performance. Confirmation comes in the form of a certificate that a firm has met a certain standard of environmental management. Standards have developed at a national level within the EU – for example the British standard – as well as at the supranational level, where work on International Standard Organisation (ISO) eco-management criteria preceded the adoption of an EU EMAS.

Taschner's analysis (see Chapter 10) makes clear, however, that NGOs remain somewhat ambivalent about the merits of EMAS. She cautions that ecoaudits often include a myriad of loopholes and that even when these can be remedied EMAS must not provide an excuse for the dismantling of more demanding forms of environmental regulation. As with ecolabel schemes, NGOs are also critical of the process whereby EMAS criteria develop within standardisation bodies which lack transparency and are dominated by industry.

Here again, instrument design and accommodating diversity within the EU and global trade context emerge as the two problematic issues. In terms of design, there has been considerable disagreement over which 'experts' are qualified to perform the tasks of verifying and certifying industrial compliance with ecoaudit standards. While independent certifiers who maintain an arm's-length relationship with industry might have incentives for more rigorous oversight, they could lack the technical sector specific expertise to perform their task properly, whereas certification by a self-administering industrial body might provide greater technical competence but runs a high risk of regulatory capture and clientilism.

Diversity among the various member state ecoaudit programmes, and the attendant risk of market distortion and consumer confusion, generated a demand for some form of coordination. Options include widespread use of a single national standard, universal adoption of the EU's own EMAS or a consensus in favour of a global ISO standard. Environmental groups favour the second option because ISO standards are considerably less demanding than EMAS (see Chapter 10). Within the EU, states have disagreed over the merits of worldwide harmonisation on the basis of a single ISO ecoaudit standard, the superior environmental characteristics versus the potential competitive disadvantages flowing from EMAS, and the legality of strict EU ecoaudit standards under GATT/WTO rules. For the moment, therefore, a fourth option has prevailed: in response to industry demands, EMAS regulation was altered to allow certification of sites which met national or international standards deemed to 'correspond' with EMAS. This requires 'bridging' the differences between the two and has resulted in serious problems of demonstrating equivalence between different ecoaudit standards.

In terms of effectiveness, ecoaudits might have illuminated some cost-cutting opportunities for firms, but have these firms fulfilled their promise of steady environmental improvement? Generalisations are difficult because

EMAS does not yet enjoy widespread use, so any environmental improvements that have resulted from this instrument are certainly not pervasive throughout the EU, while many national schemes are too recent to assess their effects. Nevertheless, available evidence suggests considerable room for improvement. In fact, as of 1997 firms in Germany comprised a staggering 70 per cent of all the EMAS registrations given out since 1993. Moreover, nearly all the firms which sought certification already met the standards, so even where it has been used it is questionable whether EMAS has exerted pressure for steady improvement.

One factor limiting the environmental effectiveness of current ecoaudit schemes, as well as their legitimacy, is a lack of transparency. Public statements required from firms under the terms of EMAS are not very demanding, making it difficult for interested parties to gauge compliance with environmental targets or the scope for possible industrial improvement. Encouraging firms to reveal additional information has not been easy because of the serious risk that it could be used against them; transparency increases the chances of being found liable for violating current or previous environmental laws. Indeed, Taschner argues that EMAS is only valuable if it is accompanied by other regulatory instruments, among which she includes liability rules. It is perhaps not surprising, therefore, that EMAS statements often fail even to provide the required information. Evidence from Germany, Belgium and Spain also suggests that ecoaudits create widespread disincentives for firms to reveal information because doing so can lead to prosecution (Chapters 3, 5 and 6). Alert to the tension between information provision and self-incrimination, environmental NGOs, industrial groups and local authorities are now faced with the broader question of whether ecoaudit schemes require or preclude liability regimes. The issue is ever more pressing in countries where liability and insurance schemes are themselves becoming increasingly popular instruments of environmental protection (Spain and Germany).

Tradeable permits

Much of the literature on new environmental instruments focuses on systems of tradeable permits which establish competitive markets among firms for emissions 'credits' originally allocated by the government. Because firms have strong incentives to reduce their pollution levels and sell excess credits to less efficient competitors, tradeable permit systems can reduce the overall compliance costs of achieving a given level of environmental protection and induce technological development which facilitates steadily greener production. Compared to the US, where permit systems have been widely used with considerable economic success (Dudek and Willey 1994; Hahn 1995), the EU has limited experience with this type of new instrument and its advantages have yet to materialise. Cremer and Fisahn cite one of the few available examples, where the poor design of Germany's permit system for air pollution undermined its environmental success and monitoring costs were found to equal those of a command and control approach (see Chapter 3). The UK has a de facto tradeable permit system for reducing SO_2 emissions but it has yet to be formalised; until it is, no clear assessments of its effectiveness are possible (see Chapter 2). Beyond the national level, various types of tradeable permit systems are currently under consideration, including ones which would incorporate Asian states (Heller, 1998).

Situating EU developments in their global context

This volume reveals that new instruments have found only moderate use within the EU, and have achieved relatively limited economic and environmental results. Nevertheless, it would be misleading to characterise the EU's supranational institutions – and certainly the fifteen member states – as international laggards in policy innovation. In fact, viewed in a global context against the records of advanced industrialised states such as the US and Japan, the EU experience with new environmental tools appears more impressive.

The US definately leads in the use of certain instruments such as tradeable permits, which were introduced in the 1970s and expanded in the 1980s, but in general its environmental policies remain dominated by a traditional command and control approach based on technology standards. Even with permits, while schemes for SO_2, volatile organic compounds (VOC), carbon monoxide (CO), nitrogen oxides (NO_x) and lead reduction constitute notable success stories, 'outside the air-pollution field, there are virtually no serious examples of decentralised, market-like approaches to pollution control that are in actual operation' (Burtraw and Portney 1991: 301; see also Hahn 1995). On green taxes, the US has hardly distinguished itself as an international standard-bearer despite highly publicised Congressional reports in 1989 and 1991 supporting their widespread use (Freeman 1994). Some individual US states have implemented taxes on various forms of waste disposal, as well as deposit-refund programmes to encourage recycling, but similar policies are found at least as frequently in Europe. Moreover, environmental levies and incentive taxes are often higher in EU member states than in the US (Hahn 1995: 145), and the proportion of total revenue derived from green taxes has been consistently much lower in the US than in most EU states (EEA 1996: 51). Besides taxes and permits, the United States employs a broad portfolio of voluntary agreements with industry, most of which are run by the federal government and the Environmental Protection Agency (EPA), but the stringency and achievements of these programmes fall short of voluntary instruments in the EU. The modesty of their economic or environmental results stems from an absence of several design features identified as essential by Biekart (see Chapter 8) – few US agreements contain actual targets and none are legally binding or backed by a strong regulatory threat (Storey 1996).

In contrast to both the EU and the US, apart from a large number of voluntary agreements at the local level which have contributed to environmental protection (Rehbinder 1994; Storey 1996), new instruments have made almost no appearance in Japan. Rather, since its belated and symbolic inception in the early 1970s, Japanese environmental policy has relied almost exclusively on technological solutions mandated through command and control regulation (Tsuru and Weidner 1989; Maull 1992). The stringency of these emissions standards and quality objectives has increased rapidly, leading to substantial environmental improvements, and the accompanying Japanese abatement technologies governing air and waste pollution are now some of the most advanced in the world (Vogel 1993). However, unlike most other highly industrialised states, where the limitations of command and control have at least framed discussions of environmental policy reform and have often led to an actual broadening of the range of environmental instruments employed, for the most part Japanese authorities have retained an 'uncritical obsession' with their exclusively technocratic approach (Meves 1992: 177). Environmentalist proposals in 1992 to establish a system of ecotaxes, for instance, encountered fierce opposition and were 'relegated [by the government] to the status of a very general discussion paper' (*ibid.*: 176).

International comparison places the EU record in a better light, but also raises an important question not directly addressed in this book: what explains international variation in instrument choice? The extent to which individual countries have departed from command and control as the dominant form of environmental policy probably depends upon a wide range of factors, including the preferences of environmental NGOs, the attitude of industrial groups, the influence of individual politicians and scientists, the position of nations within larger organisations (particularly the EU) and the institutional structure of the state. The latter presents a particularly intriguing area for further study. While institutional considerations play a secondary role in the overall extent of a nation's environmental regulation, the contributions to this volume and the record of countries such as the US and Japan suggest that they might play a more substantial role in explaining cross-national variation in instrument selection.[12] All states have retained command and control as their primary framework, but the relatively widespread adoption of new environmental instruments in Germany, Belgium and the US (and their relative absence in Japan and Britain) could reflect the scope afforded by federalism for experimentation at the regional and local level.

Conclusions

Does experience support the case for widespread use of new environmental instruments? Has their success distinguished them as viable alternatives to traditional forms of regulation? The chapters in this volume draw attention to the risks of answering either of these questions prematurely and indicate some of the fundamental considerations which will guide environmental policy making in the EU through a difficult time of transition.

With little evidence of governments actually dismantling command and control, it is methodologically difficult to determine the full benefits of new tools or their dependency on other instruments. For example, the much publicised shift to unleaded petrol, usually attributed to the effects of green taxes, was accompanied throughout Europe by command and control requirements to equip cars with catalytic converters. These devices require the use of unleaded petrol and contributed enormously to changing consumption patterns. Similarly, decreasing water pollution and improved sewage treatment in Germany and the Netherlands has resulted as much from command and control rules as from the use of taxes. Like green taxes, other new instruments have almost invariably been applied in the EU as merely one tool within a package, supplementing pre-existing command and control regulation. Without counterfactuals, which are difficult to construct and rarely offered by proponents of deregulation, one is hard pressed to conclude that, in practice, new instruments are actually superior to traditional environmental policies. In some cases new instruments have actually retarded environmental progress which might have been made through traditional command and control mechanisms (see Chapter 8).

One must also consider the argument that heavy reliance on new instruments could undermine society's quest for sufficient environmental protection by 'locking in' the wrong philosophical approach – one which worships at the altar of free-market forces (Eckersley 1993; 1995b: 12). To take one example from Eiderström (see Chapter 9), ecolabels reward greener brands, but they also legitimate rather than discourage consumption. Green consumerism by itself will not fulfil the promises made in the Rio Summit's Agenda 21 for a fundamental reduction in global resource consumption. What is needed, this argument suggests, is an approach guided by pragmatism rather than deregulatory zeal, for market instruments require a 'sustainable ecological context in which [the market's] virtue, efficiency, can shine' (Daly 1993: 182). Otherwise, without sufficiently ambitious environmental targets set by government regulation, 'an efficient servant will become an unjust and unsustainable master' (*ibid.*). This context is currently lacking in the member states – German and Dutch environmental targets are some of the toughest in the EU, but even these do not reflect sustainability objectives (see Chapter 3; see also Collier and Golub 1997).

Even if appropriate targets were devised, would a commitment to new instruments make environmental policy a hostage of government procrastination and industry hypocrisy? As many of the chapters in this volume point out, industry frequently praises new tools in theory as part of an effort to remove regulation, but then in practice opposes their adoption. Proposals for UK road taxes have languished, as have many national energy taxes, while in the Netherlands planned use of voluntary agreements has postponed the possibility of more concrete environmental measures by 8–10 years.

All of this suggests that we need packages of tools, new and old – but in which combination? While there is general agreement that policy-makers must combine the incentives and technology-inducing aspects of new instruments with direct regulation guaranteeing information and transparency (Heyvaert 1997), in many cases the appropriate mix of tools remains unclear: how, for instance, can elements of command and control such as BAT possibly coexist with taxes or tradeable permits without sacrificing their flexibility and efficiency? Recent EU framework directives on air and water pollution and on integrated pollution prevention (IPPC) illustrate this problem. Each of these laws sought to establish common environmental targets across Europe while leaving member states free to select efficient means of achieving pollution reduction. But this flexibility was simultaneously undermined by the Parliament's (and Commission's) inclusion of BAT provisions in the proposals, a move which drew considerable resistance from industry and resulted in heated debate over alternative and more vaguely worded provisions (*European Environment* 8 October 1996, 11 June 1996, 31 May 1996).[13] Moreover, there are discouraging signs that appropriate environmental policy packages will only become more difficult to fashion, as some states curtail access to information and enforcement mechanisms (see Chapter 3).

One of the most challenging implications to emerge from this volume is the need to reconsider how we judge the 'effectiveness' of new instruments, which in turn raises important questions about proper environmental policy design. While the staunchest advocates promise simultaneous cost savings, environmental improvement and political legitimacy from the arsenal of new instruments, in fact these might represent conflicting goals which trade off against each other and, not surprisingly, are prioritised and championed differently by industry, government and green groups. The three chapters written by representatives of green NGOs (Chapters 8, 9 and 10), for example, paint a rather pessimistic picture of the environmental gains actually achieved so far by ecolabels, voluntary agreements and ecoaudits. They suggest that these instruments appeal to and enjoy legitimacy within industry partly because they have been designed in a manner which emphasises flexibility and compliance-cost reduction at the expense of ambitious pollution control. Presumably, in a world where new instruments conformed to the ideals of environmental NGOs it would not be unlikely that one would find analogous accounts written by industry representatives deploring an insensitivity to cost cutting considerations.

Incompatibility between the three aspects of 'effectiveness' is a recurring theme which policy makers will have to address when weighing the respective benefits of new and old regulatory tools. For example, even if one acknowledges their economic merits (which in most cases are highly controversial), new instruments can be as anti-democratic as command and control measures (Dryzek 1995), as demonstrated by the Dutch water tax which excluded from negotiations those primarily affected (see Chapter 4). On the other hand, while the Dutch decentralised regional approach provides greater political legitimacy, by expanding the range of actors involved it creates serious collective action problems which might reduce efficiency, limit cost savings and prevent environmental gains.

The development of new instruments, either alongside or as a replacement for command and control measures, also has profound implications for the nature of governance, as it requires balancing the advantages of national diversity with the need for uniform EU rules. In some cases tailoring ecotaxes, voluntary agreements and ecolabels to local conditions can improve both the efficiency and legitimacy of environmental policy. But in other cases, as the dissenting opinion in the Molitor report notes and Commission officials discuss in this volume, excessive EU deregulation aimed at facilitating national flexibility with new instruments merely shifts environmental problems to the national level and generates coordination problems, including serious disruption of the common market when national measures constitute trade barriers (EC 1995, part 12: 16).[14]

In fact, the governance issue extends beyond the EU's borders and raises important questions about the increasing interdependence of environmental and trade policies: even if it were possible to strike an appropriate balance between member state and EU authority, can a shift towards second generation instruments be reconciled with the maintenance of global free trade governed by GATT/WTO standards (Golub, 1998b; EC 1996b)? If not, as ongoing disputes over ecolabels, ecoaudits and green taxes would seem to suggest, the EU will have to consider abandoning one of these objectives if it cannot muster political support for reforming global trade rules.

Notes

1 BAT was included, for example, in Directive 83/513 on cadmium, and Directive 82/176 on mercury discharges. In other cases EC law required only the application of best available technology not entailing excessive costs (BATNEEC), for example Directive 84/360 on air pollution. While many EC laws reflected this type of command and control approach, it would be an overstatement to suggest that they produced complete regulatory convergence across member states. The nature of EC directives, as well as certain political considerations, has provided states with a measure of discretion when setting BAT and BATNEEC standards.

2 The theoretical underpinnings of the capture model were developed by Stigler (1971) and generalised by Peltzman (1976). For evidence of capture in various policy sectors, including the environment, see Francis (1993), Bishop *et al.* (1995), McCormick (1991).
3 The deregulation debate in Europe and the US, and its relation to environmental policy, is also discussed in the two accompanying volumes in this series (Golub, 1998b; Collier 1997).
4 Many aspects of the debate over new environmental instruments stem from the seminal work of economists such as Coase (1960) and Pigou (1920).
5 Some authors in the public choice tradition have suggested that firms prefer direct regulation to new instruments because it serves as a barrier to market entry and therefore results in higher profits (Buchanan and Tullock 1975). Similarly, large firms might prefer stringent environmental regulations to block the entry of smaller competitors (Grant 1997).
6 A dissenting opinion which deplored the Molitor report's treatment of environmental issues basically as obstacles to economic activity was expressed by Pierre Carniti, a Member of the European Parliament (MEP) and former general secretary of the Italian Confederation of Free Labour Unions, and Goran Johnsson, president of the Swedish Metalworkers' Union (EC 1995: 16).
7 Markets invariably fail to provide public goods, neglect negative externalities in prices and encourage myopic planning horizons. For a discussion of these and other shortcomings, see Panayotou (1993).
8 Several of the chapters in the other volumes of this series deal specifically with economic competitiveness and new environmental instruments (Heller, 1998; Vogel, 1998; Porta 1997).
9 Problems related to calculating baselines, or 'benchmarking', also arise with green tax schemes (OECD 1994a) and generally make it difficult to compare the effectiveness of new instruments with that of command and control, a methodological point which is discussed in more detail at the end of this chapter.
10 One interesting solution was a 1991 Belgian policy which allowed a tribunal to suspend misleading environmental publicity, including misleading environmental information on labels (see Chapter 5).
11 Eiderström argues that states should not be forced to wait for EU consensus on LCA, although she expects to see a long-run convergence on criteria.
12 In his study of Great Britain, the US and Japan, Vogel finds that varying intensity of public opinion best explains cross-national differences in the extent and timing of environmental policy (Vogel 1993).
13 In the US, federal air pollution laws which require the application of 'maximally achievable control technology' have created similar impediments for flexible environmental instruments at the state level (Burtraw and Portney 1991).
14 In many instances the Commission approves new national environmental measures which limit trade or include elements of state aid, but many cases are highly contentious. For examples showing the potential trade distorting effects of new instruments, including some of those discussed in this volume, see the following issues of *European Environment*: 8 October 1996, 23 July 1996, 23 January 1996, 13 June 1995.

References

Anderson, M. (1994) *Governance by Green Taxes: Making Pollution Prevention Pay* (Manchester: Manchester University Press).

Anderson, T. and Leal, D. (1991) *Free Market Environmentalism* (San Francisco, CA: Pacific Research Institute for Public Policy).

Baggott, R. (1986) 'By voluntary agreement: the politics of instrument selection', *Public Administration* 64: 51–67.

Bishop, M., Kay, J. and Mayer, C. (eds) (1995) *The Regulatory Challenge* (Oxford: Oxford University Press).

Buchanan, J. and Tullock, G. (1975) 'Polluters' profits and political response: direct controls versus taxes', *American Economic Review* 65: 139–47.

Burtraw, D. and Portney, P. (1991) 'Environmental policy in the United States', in D. Helm (ed.) *Economic Policy Towards the Environment* (Oxford: Blackwell).

Carraro, C. and Galeotti, M. (1995) 'Economic growth, international competitiveness and environmental protection. R&D and innovation strategies with the WARM model', Fondazione ENI Enrico Mattei (FEEM), Nota di Lavoro 29.95 (Milan: FEEM).

Carraro, C. and D. Siniscalco (1993) 'Environmental policy reconsidered: the role of technological innovation', Fondazione ENI Enrico Mattei (FEEM), Nota di Lavoro 46.93 (Milan: FEEM).

Clement, K. (1995) 'Investing in Europe: government support for environmental technology', *Greener Management International* 9: 41–51.

Coase, R. (1960) 'The problem of social costs', *Journal of Law and Economics* 3(3): 215–38.

Collier, U. (ed.) (1997) *Deregulation in the European Union: Environmental Perspectives* (London: Routledge).

Collier, U. and Golub, J. (1997) 'Environmental policy and politics', in M. Rhodes, P. Heywood and V. Wright (eds) *Developments in West European Politics* (London: Macmillan).

Daly, H. (1993) 'Free-market environmentalism: turning a good servant into a bad master', *Critical Review* 6(2–3): 171–83.

Dryzek, J. (1995) 'Democracy and environmental policy instruments', in R. Eckersley (ed.) *Markets, the State and the Environment: Towards Integration* (London: Macmillan).

Dudek, D. and Willey, W. (1994) 'An overview of taxes and trading as environmental controls', in O. Hohmeyer and R. Ottinger (eds) *Social Costs of Energy: Present Status and Future Trends* (Berlin and New York: Springer-Verlag).

EC (1992) *Communication on Industrial Competitiveness and Environmental Protection*, SEC(92)1986, 4 November (Brussels: Office for Official Publications of the European Communities).

—— (1993) *Towards Sustainability*, Fifth Environmental Action Programme, (Luxembourg: Commission of the European Communities).

—— (1994) *Communication on Economic Growth and the Environment*, COM(94)465, 3 November (Brussels: Office for Official Publications of the European Communities).

—— (1995) *Report of the Group of Independent Experts on Legislative and Administrative Simplification*, COM(95)288, 21 June (Brussels: Commission of the European Communities).

—— (1996a) *Communication from the Commission to the Council and the European Parliament on Environmental Agreements*, COM(96)561final, 27 November (Brussels: Office for Official Publications of the European Communities).

—— (1996b) *European Commission Communication on Trade and the Environment*, COM(96)54, 28 February (Brussels: Office for Official Publications of the European Communities).

—— (1996c) 'Economic incentives and disincentives for environmental protection', *Proceedings of the European Commission and Council Presidency Conference*, Rome, 7 June.

Eckersley, R. (1993) 'Free market environmentalism: friend or foe?', *Environmental Politics* 2(1): 1–19.

—— (ed.) (1995a) *Markets, the State and the Environment: Towards Integration* (London: Macmillan).

—— (1995b) 'Markets, the state and the environment: an overview', in R. Eckersley (ed.) *Markets, the State and the Environment: Towards Integration* (London: Macmillan).

EEA (1996) *Environmental Taxes: Implementation and Environmental Effectiveness* (Copenhagen: European Environmental Agency).

EEB (1995) *Appeal to the European Council in Cannes: The Molitor Group Report Should be Rejected*, June (Brussels: European Environmental Bureau).

EP (1994) *Report of the Committee on the Environment, Public Health and Consumer Protection on the Need to Assess the True Costs to the Community of 'Non-environment'*, European Parliament Document A3–0112/94, 23 February, European Parliament.

Esty, D. (1994) *Greening the GATT. Trade, Environment, and the Future* (Washington, DC: Institute for International Economics).

Francis, J. (1993) *The Politics of Regulation. A Comparative Perspective* (Oxford: Blackwell).

Freeman, A. (1994) 'Economics, incentives and environmental regulation', in N. Vig and M. Kraft (eds) *Environmental Policy in the 1990s* (Washington DC: Congressional Quarterly Press).

Golub, J. (1996a) 'British sovereignty and the development of EC environmental policy', *Environmental Politics* 5(4): 700–28.

—— (1996b) 'Sovereignty and subsidiarity in EU environmental policy', *Political Studies* 44(4): 686–703.

—— (1998a) 'Global competition and EU environmental policy: introduction and overview', in J. Golub (ed.) *Global Competition and EU Environmental Policy* (London: Routledge).

—— (ed.) (1998b) *Global Competition and EU Environmental Policy* (London: Routledge).

Grabitz, E. and Zacker, C. (1989) 'Scope for action by the EC member states for the improvement of environmental protection under EEC law: the example of environmental taxes and subsidies', *Common Market Law Review* 26: 423–47.

Grant, W. (1997) 'Large firms, SMEs and European environmental deregulation', in U. Collier (ed.) *Deregulation in the European Union: Environmental Perspectives* (London: Routledge).

Hahn, R. (1993) 'Getting more environmental protection for less money: a practitioner's guide', *Oxford Review of Economic Policy* 9(9): 112–23.

—— (1995) 'Economic prescriptions for environmental problems: lessons from the United States and continental Europe', in R. Eckersley (ed.) *Markets, the State and the Environment: Towards Integration* (London: Macmillan).

Heller, T. (1998) 'The path to EU climate change policy', in J. Golub (ed.) *Global Competition and EU Environmental Policy* (London: Routledge).
Helm, D. and Pearce, D. (1990) 'Assessment: economic policy towards the environment', *Oxford Review of Economic Policy* 6(1): 1–16.
Heyvaert, V. (1997) 'Access to information in a deregulated environment', in U. Collier (ed.) *Deregulation in the European Union: Environmental Perspectives* (London: Routledge).
House of Lords (1993) *Remedying Environmental Damage*, House of Lords Select Committee on the EC, 3rd Report, 1993–4.
Jacobs, M. (1995) 'Sustainability and "the market": a typology of environmental economics', in R. Eckersley (ed.) *Markets, the State and the Environment: Towards Integration* (London: Macmillan).
La Spina, A. and Sciortino, G. (1993) 'Common agenda, southern rules: European integration and environmental change in the Mediterranean states', in D. Liefferink, P. Lowe and A. Mol (eds) *European Integration and Environmental Policy* (London: Belhaven).
McCormick, J. (1991) *British Politics and the Environment* (London: Earthscan).
Maull, H. (1992) 'Japan's global environmental policies', in A. Hurrell and B. Kingsbury (eds) *The International Politics of the Environment* (Oxford: Clarendon Press).
Meves, H. (1992) 'Japanese environmental policy: alternating stimulus and abstinence', *Japanstudien* 4: 155–82.
OECD (1975) *The Polluter Pays Principle: Definition, Analysis, Implementation* (Paris: OECD).
—— (1989) *Economic Instruments for Environmental Protection* (Paris: OECD).
—— (1990) *Financial Assistance Systems for Pollution Prevention and Control in OECD Countries*, Environment Directorate Monograph No. 33 (Paris: OECD).
—— (1994a) *The Distributive Effects of Economic Instruments for Environmental Policy* (Paris: OECD).
—— (1994b) *Applying Economic Instruments to Environmental Policies in OECD and Dynamic Non-member Economies* (Paris: OECD).
Panayotou, T. (1993) *Green Markets: The Economics of Sustainable Development* (San Francisco, CA: Institute for Contemporary Studies).
Peltzman, S. (1976) 'Toward a more general theory of regulation', *Journal of Law and Economics* 19: 211–40.
Pigou, A. (1920) *The Economics of Welfare* (London: Macmillan).
Porta, G. (1997) 'Industry and environmental policy instruments in a deregulatory climate: the business perspective', in U. Collier (ed.) *Deregulation in the European Union: Environmental Perspectives* (London: Routledge).
Rehbinder, E. (1994) 'Ecological contracts: agreements between polluters and local communities', in G. Teubner, L. Farmer and D. Murphy (eds) *The Concept and Practice of Ecological Self-organisation* (London: John Wiley & Sons).
Smith, F. and Jeffreys, K. (1993) 'A Free-market environmental vision', in D. Boaz and E. Crane (eds) *Market Liberalism: A Paradigm for the 21st Century* (Washington, DC: Cato Institute).
Smith, S. (1995) *Green Taxes and Charges: Policy and Practice in Britain and Germany* (London: Institute for Fiscal Studies).
Stigler, G. (1971) 'The theory of economic regulation', *Bell Journal of Economics and Management* 2: 3–21.

Storey, M. (1996) 'Voluntary agreements with industry', paper presented at the international workshop on the Economics and Law of Voluntary Agreements, Fondazione ENI Enrico Mattei and École des Mines de Paris, Venice, Italy, 18–19 November.

Teubner, G., L. Farmer and D. Murphy (eds) (1994) *The Concept and Practice of Ecological Self-organisation* (London: John Wiley & Sons).

Tietenberg, T. (1988) *Environmental and Natural Resource Economics*, 2nd edn (Glenville, IL: Scott Foresman).

Tsuru, S. and Weidner, H. (eds) (1989) *Environmental Policy in Japan* (Berlin: Edition Sigma).

UNICE (1993) Letter to G. Ravasio, director-general DG II, from Z. Tyszkiewicz, secretary-general of the Union of Industrial and Employers' Confederation of Europe, regarding a Community CO_2 Strategy, 8 November (on file with author).

van Goethem, A. (1992) 'The European ecolabel', *Europe Environment*, supplement to issue No. 400 (Brussels: Europe Information Service).

Vogel, D. (1993) 'Representing diffuse interests in environmental policymaking', in R. Weaver and B. Rockman (eds) *Do Institutions Matter?* (Washington DC: Brookings Institution).

—— (1998) 'EU environmental policy and the GATT/WTO', in J. Golub (ed.) *Global Competition and EU Environmental Policy* (London: Routledge).

von Weizacker, E. (1990) 'Regulatory reform and the environment: the cause for environmental taxes', in G. Majone (ed.) *Deregulation or Reregulation? Regulatory Reform in Europe and the United States* (London: Pinter).

Weale, A. (1992) *The New Politics of Pollution* (Manchester: Manchester University Press).

—— (1993) 'Nature versus the state? Markets, states, and environmental protection', *Critical Review* 6(2–3): 153–70.

Part I

NEW INSTRUMENTS IN THE MEMBER STATES

2

NEW ENVIRONMENTAL POLICY INSTRUMENTS IN THE UK

Stephen Tindale and Chris Hewett

British economists were prominent in developing the intellectual case for environmental taxation in the early years of the century, and again in urging the shift from 'command and control' to market mechanisms in the 1970s and 1980s. Yet few green taxes have actually been implemented in Britain; Britain is lagging well behind the more advanced and adventurous countries of northern Europe. Other instruments – voluntary agreements, tradeable permits, ecolabels – have also been notable mainly for their absence. This is mainly due to the relative indifference towards environmental issues among British politicians and decision-makers – there is a shortage of *any* instruments of environmental protection, new or old. However, the situation also reflects the original suspicion among many of the UK's leading environmentalists of market-based instruments, which were seen as compromising ecological integrity by placing a monetary value on environmental quality, and as offering less certainty than regulation.

The regulatory approach

As in most countries, the environmental protection regime in Britain has traditionally been based on regulation. The use of coal in London was first restricted in 1228 and the first regulation of sewers was in 1531. Regulation of water pollution was systematised in a series of Acts in 1847–8. The Alkali Act of 1863 required cuts in noxious emissions of 95 per cent; the second Alkali Act of 1874 introduced for the first time the concept on best practicable means to abate emissions and also contained the first statutory emission limit – for hydrogen chloride. The 1956 Clean Air Act – a belated reaction to the infamous London smog – gave local authorities powers to control smoke and other emissions from domestic sources. The Control of Pollution Act 1974 updated the regulations regarding air and water pollution (NSCA 1994). The 1989 Water Act created a new statutory framework for water pollution and created the National Rivers Authority (NRA) to implement it. This was the first time the job of regulator had been separated from the suppliers, and this was a result of pressure from opposition political parties and environmental groups. The 1990 Environmental Protection Act introduced the concept of integrated pollution control (IPC), under which a single regulator controlled pollution to air, water and land. The 1995 Environment Act created an Environment Agency, bringing together the NRA, the Pollution Inspectorate and the waste regulation functions previously carried out by local authorities. The most recent prominent innovation in UK environmental policy, therefore, has been to strengthen the machinery of regulation. In the climate of deregulatory zeal which dominates UK politics, this must be counted a considerable triumph for environmentalists and their allies in the civil service and government. The opposition parties had long supported the creation of a single environment regulator.

The emergence of 'new' instruments

Talk of alternatives to regulation has always been present in environmental debate, but it came to the fore and began to have an impact on policy practice only in the 1980s. Even in recent years there has been considerably more talk than action, and where instruments such as green taxes or voluntary agreements have been used they have not replaced existing regulations but, rather, been additional policy tools.

Green taxes and subsidies

It is in the field of market mechanisms that there have been some significant policy measures in recent years. A number of British airports charge higher landing fees for noisier aircraft. In the fiscal area, a price differential of just under 1 pence per litre between leaded and unleaded petrol was introduced in 1987; this has increased gradually to just under 5 pence per litre. The government has recognised the scope for increasing petrol prices, both to encourage fuel efficiency and to raise revenue. The then chancellor, Norman Lamont, announced in the March 1993 Budget that fuel duties would increase in real terms by 3 per cent in every subsequent budget. No cut-off point was given. His successor, Kenneth Clarke, increased this 'escalator' to 5 per cent in November 1993. The nearly eleected Labour government increased it to 6 per cent in July 1997.

Lamont also announced that value-added tax (VAT) would be extended to domestic energy, first at 8 per cent and then at the full rate of 17.5 per cent. He argued that this would help deliver the government's commitments to reduce carbon dioxide (CO_2) emissions. Following a sustained and high-profile campaign, the government was defeated in Parliament on its attempt to raise the rate from 8 per cent to 17.5 per cent.

From 1996 there has been a tax on landfilled waste of £7 per tonne, with a lower rate of £2 per tonne for inert wastes. About 70 per cent of all controlled waste (i.e. excluding agricultural waste and mining spoils) is currently sent to landfill; the figure for household waste is 90 per cent. Waste disposal companies will be able to avoid up to 20 per cent of their liability by paying money into special environmental trusts. This is the first new tax introduced specifically for environmental reasons in the UK.

In the 1995 Budget the chancellor announced that he was considering a number of other environmental tax changes, including creating an incentive to use gas-powered vehicles and making vehicle excise duty a banded tax to reflect a vehicle's impact on the environment. Vehicle excise duty, currently a flat rate of £140 per year for cars, could be reformed to reflect environmental impacts, with lower rates of tax for smaller, less polluting vehicles.

Since the 1989 Water Act a system of discharge 'consents' has been operated by the NRA. Companies pay for the right to discharge pollutants into water or into the air. Her Majesty's Inspectorate of Pollution (HMIP) operates a similar system with regard to IPC. However, the fees are set purely on the basis of recovering the administrative costs of the regulator and are not directly related to the volume of pollution. Industry therefore views the charges as paying for the 'regulatory service' rather than internalising external costs of their processes. There is no sense that reducing pollution would necessarily reduce the charges paid to the regulator. They cannot therefore be described as green taxes, although they could relatively easily be converted into taxes (Smith 1995).

Road pricing and tolling

The government stated for a number of years that it was considering introducing motorway tolling – most generally in the context of 'privatising roads' – that is to say, allowing a private company to build a new road and then recoup the cost through charging. However, there are obvious difficulties with this approach: Britain is a small country with a good network of non-motorway trunk roads, so motorway tolls introduced in isolation would divert a great deal of traffic on to less suitable roads, with damaging environmental consequences. The proposals also aroused considerable political opposition and the government could hardly be said to be proceeding with alacrity. In its 1996 White Paper on Transport, some years after the government had first floated the idea, it had still only got as far as 'plans for trials of electronic tolling on motorways' (Department of Transport 1996: 54).

Urban road pricing is also on the agenda, but is not making much headway. The government sponsored research, and, though some of the technical problems were held up as obstacles (despite the successful introduction of road pricing elsewhere in the world), the White Paper stated that 'the work confirmed the Government's view that price signals are a highly efficient way of influencing demand for transport' (Department of Transport 1996: 55). However, it had no intention of introducing road pricing itself and attracting the opprobrium which it assumed would follow (there is no evidence to support this view, but it is almost universally held). Instead, it planned to 'discuss with the Local Authority Associations the case for taking the necessary legislative powers to enable interested local authorities to implement experimental schemes' (*ibid.*). A more non-committal formulation is hard to imagine. For their part, even the most progressive local-authority leaders are wary of introducing road pricing, again because they assume it would cost them votes.

Subsidies

The other side of the coin is environmental subsidy, and there have been examples of this too, even from a government under pressure on public spending and committed to cutting taxes. There have been a number of schemes to subsidise householders wishing to install energy-saving measures: the latest is the Home Energy Efficiency Scheme, which targets support for insulation at low-income households. This came from general budget expenditure; other subsidies have been funded by de facto levies on consumers, administered through the price-regulation system which operates in the UK's privatised energy industry. The Energy Savings Trust, a joint government/industry body, has subsidised low-energy light bulbs and gas condenser boilers. All of these examples, however, are very small in cash terms.

A more substantial subsidy arrangement is the fossil-fuel levy/non-fossil-fuel obligation. Consumers pay a levy on the final bill (it has been around 10 per cent) and the money is given to electricity companies to compensate them for buying fuel from 'clean' sources at above market rates, which they are obliged to do. Environmentalists' enthusiasm for this scheme is greatly tempered by the fact that over 90 per cent of the levy goes to the nuclear industry (non-fossil fuel, certainly, but hardly clean), but it has nevertheless given a very significant boost to renewable energy generation. With the introduction of liberalisation into UK energy supply, and following complaints from the European Commission, the nuclear portion of this levy is to be phased out by 1998. The support for renewable energy, however, is likely to continue with the blessing of the Commission.

Unlike many other countries, the UK does not offer low-interest loans for installing cleaner technologies. There is some support for research and development (R&D) in this area, but the amounts of money are negligible. There is also some support for small and medium-sized enterprises (SMEs), but this is limited to the provision of information on pollution reduction and 50 per cent grants for the costs entailed in registering on the EC eco-management and audit scheme (EMAS).

There is a growing lobby representing the interests of the environment industry in the UK, and it is quick to point out the many areas where other countries support clean technology and the UK does not. In particular, there are calls for tax allowances for replacing old 'dirty' technologies with new cleaner ones, export credits for UK environment-technology companies and an assessment of the tax system to attract more investment in innovative clean technologies (Wilkes 1995). As yet, however, there has been little government action in this area.

Water metering

A failed attempt to introduce a new instrument was the plan to change the basis of charging for domestic water supply from the current property-based system to volumetric charging through metering. Since the abolition of domestic rates in the late 1980s there have been no rateable values for new houses, so the current system of charging cannot be used. All new properties are therefore being fitted with meters. However, the government wished to speed up the changeover, partly to provide consumers with an incentive to use water more sparingly and so reduce the problems caused by over-abstraction. Some of the water companies – notably Anglian, which has particular problems with water shortages – attempted to move all their customers on to meters. Other companies were persuaded by the regulator to offer reduced tariffs for customers choosing meters. It is likely that increasing numbers of customers will opt for a meter. However, compulsory metering has proved extremely unpopular, and those companies which adopted it as policy have been forced to back down. The Labour Party campaigned hard against compulsory metering and its case was strengthened by evidence from national metering trials which suggested that reductions in consumption were uncertain and might not last, and that metering was extremely expensive – between £165 and £200 per property (Water Services Association 1993). A more cost-effective means of conserving water, Labour argued, would be to spend money plugging leaks in the distribution system.

Tradeable permits

The privatisation of the electricity industry and the break-up of the monopoly generator into competing companies necessitated a rethink on acid-rain regulation. Each generating company was given emissions caps for sulphur dioxide (SO_2). Companies can transfer emissions quotas between themselves. This is a de facto tradeable-permits system, though it has not been described as such, and the move to introduce more competition into generation will probably require the introduction of a more formalised system. Indeed, the government has consulted on the possibility of a sulphur trading system for the UK, building on US experience.

Exhortation and voluntary agreements

There have also been a number of attempts to secure environmental change through voluntary agreements with industry or by changing consumer behaviour (without offering a fiscal incentive). In 1991 the government launched the latest in a series of public campaigns to persuade people to use less energy in the home. Called Helping the Earth Begins at Home, the campaign featured a series of somewhat kitsch advertisements in which a child exhorted everyone to think about her future. The government argue that the campaign had some impact on increasing knowledge of the link between energy use and climate change, but did not even claim that the campaign actually had any impact on consumer behaviour. In 1994 the campaign was renamed Wasting Energy Costs the Earth and relaunched, but the new version was no more successful.

In the corporate sector, the government runs an Energy Efficiency Best Practice Programme which spreads the word about possible savings. It claims it has delivered annual reductions in carbon emissions of 2 million tonnes. The government also runs a campaign called Making a Corporate Commitment, under which companies promise to reduce energy use. By 1995 1,850 businesses had signed, but the impact appears limited: 23 per cent of signatories said they had no plans to implement energy efficiency measures in the next year, compared to 15 per cent the previous year. There were also no obligations for the companies to report progress towards their targets. Under a separate scheme, UK car manufacturers have signed up to a target of a 10 per cent improvement in vehicle efficiency by 2005.

From a policy perspective, the more interesting developments have been where government has been involved in negotiating agreements with trade associations representing whole sectors of industry. Assessments of such agreements must be case by case as the details and commitments vary widely.

The first agreements, signed by government and trade associations, have dealt with the use of hydrofluorocarbons (HFCs) in industries, including aerosol, foam and refrigeration. Neither government nor industry regard the agreements as binding. Industry promises to minimise the use of HFCs and government has declared that it will not impose reduction targets while the agreements are in place. There was very little transparency in the negotiations, with little or no involvement of non-governmental organisations (NGOs), although given that some are calling for a ban on the substances that is perhaps not surprising. More worrying is that not all the agreements have reporting requirements, and those that *do* have no requirement for independent monitoring.

A more robust set of agreements is taking shape as part of the government's waste-management strategy. Under the heading 'producer responsibility', the government is negotiating with manufacturers and sellers of products and their packaging to ensure that 'the industry assumes an increased share of the responsibility for the waster arising from the disposal of its products' (Department of the Environment 1995a: 44). The first sectors to be involved are packaging, vehicles, newspapers, tyres, batteries and electronic equipment. The first to publish targets for consultation was the packaging industry, which promised to recover 50 per cent of packaging waste by 2000. This is within the range of 50–70 per cent by 2001 set by the EC Packaging Directive.

The producer-responsibility initiative has a specific base in legislation, through the Environment Act 1995, and it appears that government has been much more explicit about threatening to use other instruments if industry does not produce results. In some cases, including packaging, industry itself has requested legislation to back up voluntary agreements to prevent the problem of free riders. There has also been discussion that the packaging agreement will incorporate a levy to help finance some of the recovery schemes, but it is still not clear whether this will be introduced. The fact that the government is now drafting regulations, of course, means it would be inaccurate to describe 'producer responsibility' as constituting voluntary agreements any more. The best way of characterising the policy instrument would be 'negotiated regulations', in as much as the industry voluntarily requested regulation.

The Chemical Industries Association has recently signed an agreement with government to reduce its members' energy consumption per tonne of product by 20 per cent between 1990 and 2005. This agreement, the first of its kind in the UK, has been hailed as groundbreaking by the government but criticised by commentators as poor on transparency and difficult to enforce (ENDS 1997).

The final area where a voluntary approach is being mooted is in the enforcement of current regulations. The pollution inspectors have floated the idea that if a company is signed up to an environmental management standard, EMAS or BS7750, there may be a reduced need for monitoring of actual emissions. A senior pollution inspector has recently described this as 'a risk-based approach to the setting of inspection and monitoring programmes' (Duncan 1996). This is undoubtedly being driven by the pressure of financial resources on the Environment Agency, and some questions are being raised about its legal basis.

Reasons for the shift to new instruments

Some of the changes were made necessary by the Conservatives' radical programme of reform. Tradeable permits, as we have just seen, were introduced as a response to privatisation and the introduction of competition. Compulsory water metering for new houses became inevitable when the government abolished domestic rates (to which the previous charging system had been tied) in favour of the poll tax (later revised as council tax). Motorway tolling, if it ever materialises, will be a result not of an attempt to manage demand but of the desire to attract private finance into road building.

Other changes have been introduced under the guise of environmental concern but have in reality been straightforward revenue raisers. Lamont's claim that VAT on fuel was part of the government's climate-change strategy was ridiculed by the Labour politician Robin Cook, who noted that it had more to do with the public finances going red than Treasury ministers going green (a consultation paper on the climate-change strategy, published just three months earlier, had not mentioned extending VAT even as a possibility). With some measures both environmental and revenue-raising motives were present (and there is nothing wrong with that – the government has to get its money from somewhere). The road-fuel escalator is an excellent environmental measure, but it is also an extremely nice earner for the Exchequer: Lamont estimated that his 3 per cent escalator would net him an extra £1 billion in the third year. Of course, some of the income would be eroded by changes in behaviour – people driving less to reduce fuel costs – but as the duties are set to increase every year, predicting the revenues is essentially a balancing act. The two objectives of raising revenue and changing behaviour need not contradict each other.

Even where structural reform or fiscal pressure has not required new environmental instruments, the pressure for a move away from command and control has not come from Britain's environmentalists, most of whom were reluctant to abandon the tried and tested role of regulation. It did not come from the opposition Labour Party, which was generally indifferent to environmental issues and frightened of opening up debate on the thorny subject of taxation. It has come partly from British industry, which pays lip service to the superiority of market mechanisms over regulation but in general opposes both, and now argues that voluntary agreements are the best way forward.

Such initiative that there is has come almost entirely from the previous Conservative government itself, supported by academics and the right-wing think-tanks which have been highly influential with the Thatcherite wing of the Conservative Party. The intellectual antecedents can be traced back to Pigou's *Economics of Welfare*, published in 1920, but the modern debate came to prominence with a minority report by the Royal Commission on

Environmental Pollution (RCEP 1972), signed by the Oxford Economist Wilfred Beckerman and the scientist Lord Zuckerman, which urged the government to introduce pollution charges. Beckerman subsequently outlined his arguments in a paper for the Institute of Economic Affairs (Beckerman 1975), a free-market body close to the then newly elected Conservative leader Margaret Thatcher. Thatcher was not much interested in environmental issues per se, but clearly her government was to be so anti-regulation that it would favour the use of market mechanisms in most areas of policy – environment included.

The free-market think-tanks retained an interest in market mechanisms for environmental protection through the 1980s and 1990s – as, for example, in the Adam Smith Institute's 1992 publication *The Market in Environment* (Taylor 1992). Their role was not wholly constructive, however. An equal amount of their research was directed at attacking the science behind some major environmental issues, particularly climate change (Bates and Morris 1994).

The issue of 'market v. regulation' was also used by the Right to further denigrate the role of the European Commission, which, at the time, was mostly concerned with regulatory mechanisms. As we will see later, however, when the Commission came to propose its own market mechanism the UK government opposed that as well.

As noted above, little was done to introduce environmental taxes. The first significant move came in 1987, with the introduction of the tax differential between leaded and unleaded petrol. This was hardly revolutionary, in the context of the boldness (or foolhardiness, depending on one's perspective) of other Conservative tax reforms. (For example, Geoffrey Howe had reduced the standard rate of income tax and doubled the rate of VAT in his first Budget.) It was essentially a reactive move; the government was under pressure from a well-organised group called the Campaign for Lead Free Air (CLEAR) and felt it had to do something. Nevertheless, it is significant that the response chosen was fiscal rather than regulatory. A tax differential fitted more comfortably into the ideological framework. It was less unacceptable to industry than the alternatives – in particular the mandatory fitting of catalytic converters (which require unleaded petrol), which the British motoring industry was strongly opposing. Environmentalists were, on the whole, happy that *something* had been done, though there was scepticism about how effective it would prove and some criticism of the fact that duty on unleaded petrol had been reduced rather than duty on leaded being increased.

After 1987 environmentalists had to wait another six years for the next significant fiscal measure. By 1993 the combination of the momentum of the post-Rio process and (more importantly) the government's fiscal crisis was enough to persuade Norman Lamont to introduce the fuel escalator and VAT on domestic fuel. Debate was not absent, however. The intellectual case for market mechanisms was given a boost by the publication of a government-sponsored report called *Blueprint for a Green Economy* (Pearce *et al.* 1989). The European Commission proposed its carbon/energy tax. The government published its first White Paper on the environment, *This Common Inheritance* (Department of the Environment 1990). In the face of all these developments, opposition parties and environmental groups were compelled to face up to the changing agenda.

Each of these developments will be examined. First, however, it is worth pointing out that the main event in British environmental politics in the period 1987–93, the 1990 Environmental Protection Act, was almost entirely concerned with the strengthening of old instruments – the air, water and soil pollution-control regimes – combining them into a single IPC system. Despite its deregulatory zeal, the government appeared to lack the courage of its convictions, fearing to dismantle environmental controls for fear of alarming the public. Indeed, the quote which appears at the front of *This Common Inheritance* seems to confirm the triumph of the old approach: 'Is there not the Earth itself, its forests and waters, above and below the surface? These are the inheritance of the human race. . . . No function of government is less optional than the *regulation* of these things' (J. S. Mill; quoted in Department of the Environment 1990; emphasis added). The quote is from John Stuart Mill, an interventionist liberal who tended in later life towards socialism – a strange choice of authority for a free-market government.

Nevertheless, *This Common Inheritance* did engage the government firmly in the debate on the merits of market mechanisms. The previous year a group of academics led by David Pearce of University College, London, published a study of environmental economics which, significantly, had been sponsored by the Department of the Environment and which they called – effectively in marketing terms – *Blueprint for a Green Economy* (Pearce *et al.* 1989). Pearce and his colleagues argued, first, that 'business as usual' was not a sustainable proposition; second, that taxes were a more efficient means of changing behaviour than regulation; and, third (and more controversially), that the proper level of such taxes could be calculated by placing a monetary value on environmental quality.

Blueprint was widely acclaimed and even more widely debated. The government, and in particular the Department of the Environment, seemed sympathetic. Yet by the time of the 1990 White Paper, enthusiasm had cooled. The paper stated that:

> Regulation [has] limitations. It can be expensive to monitor and difficult to up-date quickly in response to scientific and technical advance. It cannot always pitch controls at the level which strikes the most effective balance between environmental benefits and compliance costs. . . . Regulation has always been required and is

> still required, but it has its shortcomings. For these reasons the Government, along with other governments throughout the world, has begun to look for ways to control pollution which avoid some of these problems by working with the grain of the market.
>
> (Department of the Environment 1990: 13)

Yet most discussion of Pearce was confined to an annexe on 'Economic Instruments for Environmental Protection'. Despite numerous promises to consider this, explore that and consult on the other, there were no commitments to action.

What were the reasons for this timidity? The main factor was undoubtedly industry pressure. To the extent that generalisations can be made, it is fair to say that British industry has adopted a negative attitude towards environmental policies. Despite mounting evidence that, in the Confederation of British Industry's (CBI) phrase, 'Environment means Business' – that there is money to be saved or made in taking the environment seriously – British industry has generally regarded proposals for new environmental policies with hostility, seeing them as inevitably leading to higher costs and reduced competitiveness. Companies and their associations have accepted the general case that market mechanisms are more efficient than regulations, but have proceeded to argue against virtually every specific market mechanism that is proposed. A good example of this is BP (British Petroleum), which produced a discussion paper praising the efficiency of market mechanisms while flatly opposing the main proposal on the table (British Petroleum 1995).

This was a proposal which came not from London but from Brussels, the European Commission's proposal for a carbon/energy tax. Every effort was made to accommodate the concerns of the business sector: the tax would be fiscally neutral, with the proceeds returned to companies via lower social-security contributions; energy-intensive sectors would be exempted; border-tax adjustments would be used to prevent unfair competition from outside Europe; and, to make absolutely sure, the entire package would be made dependent on similar action being taken by Europe's main competitors. Yet still industry cried wolf – or, rather, 'damage to competitiveness', which to modern ears is even more alarming. In opposing the carbon/energy tax, industry was certainly not supporting a regulatory approach instead. What it wanted – and got – was in effect no European action to combat climate change, to be allowed to go on polluting in its merry, profligate way. (In fairness to British industry, it should be said that industry lobbies from most other member states were not much more enlightened.) Over the last two years, industry has concentrated on other policy instruments. The CBI now recommend that 'business-led, market based voluntary action should always be the first recourse of Government when seeking environmental improvement' (Confederation of British Industry 1994: 30).

The carbon/energy tax dominated debate about environmental taxes in the period up to 1993, just as VAT on fuel has dominated debate since 1993. It was the litmus test; if one opposed this specific proposal, one had to oppose green taxes generally. And the British government did oppose it, not only because its industrial backers asked it to, but also because the proposal emanated from Brussels. Even worse, it was associated with Margaret Thatcher's *bête noire*, the man trying to reintroduce socialism to the newly liberated Britain, the Commission president, Jacques Delors. Thus sensible debate about green taxes was an early victim of the extraordinary internecine warfare which broke out within the ruling Conservative Party over Europe and which has continued to this day.

If the government moved swiftly to relegate Pearce to an appendix, *Blueprint* did succeed in focusing the minds of the opposition and the environmental movement on green taxes. So too did a 1990 publication from the prestigious and non-aligned Institute for Fiscal Studies (IFS), *Taxation and Environmental Policy: Some Initial Evidence* (Pearson and Smith 1990), which again argued the superiority of taxes over regulation.

The Labour Party, in schizophrenic mode, halfway between its command-economy phase and the unquestioning neo-liberalism of 'New Labour', appeared unconvinced that taxes were better than regulations. Though its 1990 environment policy statement *An Earthly Chance* accepted the polluter-pays principle, it argued that:

> not all environmental damage can readily be costed in financial terms and billed accordingly. The 'polluter pays' principle is a means of attributing responsibility. It leaves open the question of whether it should be implemented through the price mechanism or through regulation.
>
> (Labour Party 1990: 9)

More positively, it stated that 'we are attracted to the concept of "green taxes" as a substitute for other forms of taxation so they are fiscally neutral but directed towards environmental objectives' (Labour Party 1990: 10). However, it specifically ruled out the carbon/energy tax, which, it was argued, would drive the UK into recession, reduce competitiveness and, without compensation, fall heavily on the poor.

The think-tanks of the Left, however, were attempting to emulate their right-wing counterparts by pressing the case for green taxes. The newly created Institute for Public Policy Research (IPPR), which had close links to Neil Kinnock and other Labour 'modernisers', produced a paper in favour of road pricing in 1989 (Hewitt 1989) and a 'Budget Memorandum' on green taxes in 1990 (Owens *et al.* 1990), while the Fabian Society published *Sustainable Development: Greening the Economy* (Jacobs 1990) the same year. Both challenged the view that green taxes were inherently right wing and regulation left wing. Both are based on a recognition that intervention is needed in the operation of free markets and argue that the choice of instruments should be made simply on grounds of appropriateness for particular issues.

Rethinking was also under way in environmental pressure groups, particularly one of the largest, Friends of the Earth. *Blueprint* had been greeted with scepticism bordering on hostility, particularly for its argument that a valuation could be placed on environmental quality. Regulation was seen as a tried and tested method producing secure and predictable outcomes. The argument that taxes are more cost-efficient than regulations cuts little ice with many environmentalists, who see the private sector as rapacious and extremely rich. Cost–benefit analysis is regarded with great suspicion because of a particularly perverse and indefensible form of it which has been used by the Department of Transport to justify building new roads (public open spaces, because they cannot be built on by anyone except government, are given zero value). The inclusion in the 1995 Environment Act of a requirement for the Environment Agency to take account of the costs of its proposed actions – a fairly commonsensical proposition – caused outrage among environmental lobbyists.

Nevertheless, it was recognised that taxes would have to play a role, particularly in persuading consumers to be less profligate. Friends of the Earth's contribution was to confront the issue of equity: would environmental taxes penalise the poor? Friends of the Earth commissioned the IFS to look into this, resulting in a report called *The Distributional Consequences of Environmental Taxes* (Johnson *et al.* 1990), which remains the most influential work in the area. The IFS authors argued that higher petrol prices would be progressive across the population as a whole (though regressive among car drivers), higher food prices would be somewhat regressive and higher domestic-energy prices highly regressive. They also recommended various options for compensation packages.

The various reports and research projects did succeed to an extent in changing the intellectual climate and blunting some of the previous opposition to green taxation among social policy groups, trade unions and politicians of the left. The Labour Party's 1994 environment policy document *In Trust for Tomorrow* was more open to the use of environmental taxes, arguing that they should be used to create markets for certain products, to raise money to fund the clean-up of contaminated land and to encourage the use of fuel-efficient cars and lower car use (Labour had not opposed the fuel escalator). Labour subsequently supported the landfill tax. *In Trust for Tomorrow* rejected the carbon/energy tax because it would increase domestic-energy prices – this stance was inevitable given the party's successful campaign against VAT on fuel – but stated that it might be prepared to consider a tax shift in the non-domestic sector, increasing energy prices while reducing employment taxes.

The smaller Liberal Democrat Party was (and remains) the most enthusiastic of the three main parties. This is partly because it has inherited from its predecessor, the Liberal Party, a tradition of radical locally based activist politics which is sympathetic to environmentalism, and partly because, as the name implies, it has a commitment to liberal economics, which makes it predisposed to favour market mechanisms over regulation. Indeed, an issue combining market economics with environmental concern could have been designed for the Liberal Democrats. It is also fair to say that the Liberal Democrats, a party with very few seats (though a much higher proportion of the popular vote), can afford to be much more open and radical in its policy positions than either of the main parties. Its proposals attract much less scrutiny and criticism, because everyone knows that they will not be implemented – at least not by the Liberal Democrats. The party is in the unenviable position of being a glorified think-tank or pressure group, seeing its ideas adopted by others with never a word of thanks. The fact that the Liberal Democrats have embraced environmental taxation gives some grounds for optimism that a Conservative or Labour government might move in a similar direction at some stage. This process could be greatly accelerated were a Liberal/Labour coalition to emerge.

Increased interest in voluntary agreements

More recently, the UK government has become increasingly interested in the use of 'voluntary agreements' with industry to implement environmental improvements. As discussed earlier, the term 'voluntary agreement' is used very broadly in the policy context. It can be mere exhortation to industry to use best practice or implement environmental management standards. However, the most significant situation from our perspective is where government and an industrial sector, represented by a trade association, are negotiating specific environmental targets. The government has had a policy to seek such agreements with industry since 1993 but the first of these was signed only in January 1996.

The interest in voluntary agreements is almost completely down to the influence of industry on government policy. Business has been hostile to new environmental regulations and, while supporting the principle of market mechanisms, has generally opposed specific green tax measures as well. In a number of sectors government has allowed itself to be convinced by industry that both of these types of measure will damage competitiveness. In that sense the move to voluntary agreements is very reactive and has been strongly criticised by environmental NGOs (Jenkins 1996).

One should not dismiss the initiatives out of hand, however. In a number of cases the agreements are linked to legislation, either implicitly, by government threatening to regulate the sector if progress is not made voluntarily, or explicitly, where the sector itself has called for regulation to prevent 'free riders' or a levy to support the necessary changes. Ironically, some uses of the voluntary approach, such as producer responsibility, have reinforced the arguments for regulation or fiscal measures.

Have green taxes been beneficial?

It is generally asserted that the Conservative government's first foray into green taxation, the price differential between leaded and unleaded petrol, was a resounding success. The percentage of petrol bought which was unleaded increased from less than 5 per cent to over 50 per cent at the start of 1993 (when the mandatory fitting of catalytic converters to new cars was introduced, making it impossible to attribute further increases in unleaded sales between market and regulatory measures). Lead emissions fell by half. This success is not surprising. For most motorists substitution was extremely easy, so even a small incentive was enough to change behaviour. There is a prima facie case for saying this was a successful use of a market mechanism, but it is not entirely clear that it was more effective or even more cost-efficient than a regulatory approach (mandatory catalytic converters or bans on sales of unleaded). This was a new instrument altogether, not a market instrument replacing a regulation. So any attempt to compare the two would necessarily be counterfactual. It has not, to our knowledge, been attempted.

This is true for all of the UK's implementation of 'new' policy instruments. Despite government rhetoric about deregulation and the efficiency of market mechanisms, to our knowledge no regulatory approaches have been replaced by taxes or voluntary agreements. The substitution has only occurred where there is a choice between a 'new' regulation or a new market mechanism.

Has the fuel escalator, a more significant environmental tax, been beneficial? This is again hard to judge, since the things it is trying to affect – fuel efficiency of vehicles and vehicle use – are subject to myriad other factors. Overall petrol consumption fell by 4 per cent in the first year after the escalator was introduced, but this is unlikely to have been much due to the escalator, which is *intended* as a long-term, gradualist policy. Indeed, the fall had begun in 1991. The escalator has little short-term impact on fuel prices, particularly as compared to fluctuations caused by changes in world oil prices.

This is, however, one of its great virtues. Incremental change has proved a relatively painless way of increasing environmental taxation. The escalator has united the political parties (though the Liberal Democrats are concerned about its impact on rural motorists – most of their MPs represent rural constituencies). It provides a good model for how to introduce a new tax. However, it is not clear how widely known it is, and therefore how widely the expectation of higher prices in the future is affecting model choice or location decisions. Anecdotal evidence suggests that hardly anyone is aware of it, though the manufacturers are and may be altering their plans accordingly.

The new landfill tax is also a good model of how a tax ought to be introduced. It was announced in 1994, details were announced in 1995 and extensive consultation was carried out. The original proposal from the government was for an ad valorem tax, which was easier for the Treasury to administer but did not accurately reflect environmental costs. Both the waste industry and NGOs told the government that a weight-based tax would make more sense. Faced with this near-unanimous response, the government changed its mind and a weight-based landfill tax was implemented in 1996. This is a new departure, not only for environmental policy, but for British tax policy generally. Again, all political parties have supported it, and all environmental groups. Even business has not complained too much, although there has been much lobbying about what should constitute inert waste and so be liable for a lower tax rate.

Which green taxes have failed?

VAT on fuel

Mention environmental taxation to a British politician and he or she will think of VAT on domestic fuel, shudder and change the subject. The episode is a sorry saga of political incompetence which has set back the cause of environmental taxation – and environmentalism in general – and may have helped lose the Conservatives the last election.

Environmentalists and opposition politicians may have argued, correctly, that the British government's decision to impose VAT on domestic fuel and power was motivated by financial rather than environmental considerations, but the fact remains that the other proposals for energy taxation, such as the European Commission's carbon/energy tax proposal, would have had a broadly similar impact on domestic-energy prices (albeit phased in over a longer period).

The government justified its proposal to levy VAT on domestic fuel on the grounds that this was necessary to meet the carbon-emissions reduction target agreed at the Rio Summit. However, it estimated that the new tax would cut emissions by just 1.5 million tonnes of carbon per year, less than 1 per cent of the total.

The government also argued, correctly, that most other European countries have a tax on domestic fuel, ranging from over 90 per cent for domestic electricity in Denmark to 17.5 per cent in the Netherlands. But it ignored a crucial difference. Other northern European countries with climates similar to or colder than Britain's have far stricter regulations governing the insulation standards of their housing stock. They simply do not have the draughty, damp and impossible-to-heat properties which are so common in the UK. Energy use is therefore much more closely correlated to income – those who use more energy do so because they own more appliances or indulge in more luxuries. Until Britain reaches similar levels of energy efficiency with its housing stock, comparisons with domestic-energy taxation in other northern European countries will be entirely bogus.

Public support for the concept of 'fair taxation' remains strong, as the Conservatives discovered when they attempted to introduce a poll tax – an approach which had offended medieval conceptions of social justice and was therefore unlikely to appeal in the more enlightened twentieth century. The concept of progressive taxation is supported by 85 per cent of the British public. And it soon became clear that VAT on domestic fuel would be a very unfair tax.

The social impact of higher fuel bills is illustrated by a study carried out at the University of York's Social Policy Research Unit (Hutton and Hardman 1993). The survey found that those households with incomes in the top 20 per cent spend 4.2 per cent of their budget on fuel, while those in the lowest 20 per cent spend 12.1 per cent. The burden is therefore nearly three times greater for low-income households than for more affluent households. Low-income households are less able to cut back on fuel use by changing equipment or installing energy-efficiency measures, which can have a high capital cost. Only 46 per cent of those in the lowest quintile have gas central heating (the most energy-efficient form of space heating); the figure for the highest quintile is 75 per cent.

There are also different impacts for different types of household. Families with children spend over £13 per week on fuel, while pensioners and single householders spend £8–£10 per week. Single pensioners spend 16.36 per cent of their budget on fuel, but lone parents with children under 5 spend 16.42 per cent. Those living in private rented accommodation – the most energy-inefficient form of tenure – also have proportionately high fuel bills, and less incentive to invest in efficiency measures since they may not stay in the property long enough to reap the benefits. These figures illustrate the difficulty of designing a targeted compensation package.

Once it realised the unpopularity of the proposed tax increase the government attempted to overcome the problem of regressiveness by increasing the budget of the Home Energy Efficiency Scheme, which gives grants to low-income households to install insulation and pipe-lagging. The scheme was also extended to include all pensioners; this had political advantages but reduced the funds available to low-income households. Energy-conservation groups estimated that it would take over a decade to insulate the dwellings of all low-income households – a decade in which the fuel-poor would be faced with higher bills.

The Labour Party, which had always opposed the carbon tax (the effects of which would have been similar), savagely attacked the extension of VAT –

its task made easier by the fact that the Conservatives had fought and won the 1992 General Election on a promise not to increase taxes generally and VAT in particular. The Liberal Democrats also came out against the VAT increase, even though they had supported the carbon/energy tax. Environmental groups were generally agreed that the tax should have been accompanied by regulations and other measures to increase its efficacy, and by a compensation package. But they split in terms of presentation. Friends of the Earth welcomed the tax but called for the accompanying measures to be added. Greenpeace condemned the tax but said it would support it if the other measures were introduced. This led to a farcical exchange in the House of Commons, with the prime minister quoting Friends of the Earth and the then Leader of the Labour Party, John Smith, quoting Greenpeace back.

In the event the government succeeded in imposing the tax at 8 per cent but was defeated in a vote on its attempt to increase it to 17.5 per cent. It raised virtually no revenue – most had to be conceded to pensioners and the poor in grudging compensation packages. The political price was high; government popularity plummeted and a series of spectacular by-election losses followed. The environmental price, in terms of the political feasibility of implementing other measures, was also high and cannot yet be assessed. The scope for sensible debate about environmental taxes was radically reduced. The impression of a trade-off between environmental protection and social policy, which environmentalists have been striving to overcome for a decade, was reinforced. So too was the widespread view that environmentalism equals sacrifice and so will not be popular with the voters, another myth which environmentalists have sought mightily to bury. Deservedly, in the light of all this, Norman Lamont lost his job. The new Labour Government cut VAT on fuel to 5 per cent, the lowest rate it could set under European rules.

Have voluntary agreements been a success?

There is a great deal of concern about the environmental effectiveness of voluntary agreements, and the lack of transparency and targets can only reinforce that concern. So far, the most concrete results to have come from a UK voluntary initiative seem to be from that of producer responsibility in packaging: it has clear targets and public documents. In the end, however, industry felt it could not deliver without back-up in terms of regulations and/or levies. The initiative therefore ceased to be voluntary. The targets are also taken from the EU Packaging Directive, which is mandatory for all member states. It is debatable whether the same results would have been reached without the Directive. Other agreements seem to bear out this concern. They are far less transparent and so far have no binding targets. Without these elements they are little more than warm words, and to reach any ambitious targets it appears that more than just a voluntary agreement is needed.

Prospects for new instruments

The Liberal Democrats have committed themselves not only to a number of new product taxes but also to a significant shift in taxation, involving increases in energy taxes and reductions in VAT. This concept of a fiscally neutral 'environmental tax reform' has been gaining support gradually in recent years, though it is not nearly as widely debated in the UK as in many other European countries. The Conservative chancellor Kenneth Clarke signed up to the principle when he linked the new landfill tax to a reduction in employers' National Insurance Contributions, saying: 'I want to increase the tax on polluters, and make further cuts in the tax on jobs' (Budget Statement 1994). (He also accepted the arguments that environmental taxes can make the economy work better: 'in some cases, taxes actually do some good, by helping markets work better and by discouraging harmful or wasteful activity' (*ibid.*) However, the landfill tax raises a comparatively small amount of money, so the shift cannot be said to be significant, and his 1995 Budget contained no new taxes on polluters.

A number of government advisory bodies have supported the principle of environmental tax reform. The Panel of Sustainable Development, set up by John Major after the Rio Summit to advise him on environmental matters, has called for 'wider use of economic instruments, and a gradual move away from taxes on labour, income, profits and capital towards taxes on pollution and the use of resources, including energy' (Department of the Environment 1995b: 11). The panel is chaired by Sir Crispin Tickell, former ambassador to the UN and one of Britain's leading environmentalists – he is credited with the temporary greening of Margaret Thatcher in 1988. He did not achieve the same feat, even temporarily, with John Major.

Another government-sponsored body, the Advisory Committee on Business and the Environment (ACBE), which includes representatives from major energy, construction, retailing and financial services companies, has stated that 'ACBE welcomed the Chancellor's statement in his 1994 Budget that in future the Government would be looking to shift the burden of taxation from wealth creation to resource use and pollution' (Department of Trade and Industry 1996: 49). The recently launched Real World Coalition, which brings together environmental and social policy groups (itself a significant development) strongly supports a tax shift.

The new Labour government has issued a *Statement of Intent* on environmental taxation, saying it will 'explore the scope for using the tax system to deliver environmental objectives. . . . Over time the Government will aim to reform the tax system to increase incentives to reduce environmental damage' (HM Treasury 1997a). It has followed up this statement with a further increase in the road-fuel duty escalator (to 6 per cent) and specific consultations on landfill tax, water-pollution charges and aggregates tax (HM Treasury 1997b). Before the election the left-wing think-tank IPPR attempted to push the debate forward by publishing a specific set of tax proposals (Tindale and Holtham 1996). Clearly the new government is interested in the ideas, but it is too early to predict how radical they will be. If it comes, reform is more likely to be a rolling programme of green tax measures, rather than the large packages introduced in the Scandinavian countries.

Conclusions: how should responsibility for new instruments be divided between the EU and member states?

Like any other policy instrument, the way in which market mechanisms and voluntary agreements are applied will depend on the environmental problem to be solved. One of the strengths of the debate on green taxes is the constantly growing list of examples from around Europe being implemented at member-state level and, still more, at local and regional level. Innovation in the use of market mechanisms should be encouraged, but should the EU itself take a lead?

The fate of the carbon/energy-tax proposal, and the role of the UK in that debate, does not bode well for action on green taxes at an EU level. The UK, not alone in the EU, has a very strong belief that fiscal matters should be left to the nation-state. The change in government in the UK is unlikely to alter that stance. Indeed the Labour Party opposed the carbon/energy tax and has stated it would not advocate qualified-majority voting (QMV) for fiscal instruments of any kind at EU level, including ones dealing with the environment.

The most effective role for the EU, it seems, will be to help coordinate the use of market mechanisms at member-state level. The review of mineral-oil duties is one opportunity where the EU can increase the minimum level, thus indicating to member states the direction in which energy taxes should be heading. Additional information from the Commission on how green taxes impinge on EU internal-market rules could also be used as an encouragement for member states to innovate.

The UK government recently used the Climate Change Convention to call for a global tax on aviation fuel (Department of the Environment 1996b). One stepping stone towards this goal could be for the EU to remove the tax exemption for air fuel on internal EU flights. Further removal of the exemption would need agreement at international level.

The EU may have a more important role to play in supporting voluntary agreements. It appears that the most effective voluntary initiatives with industrial sectors have been underpinned by EU framework legislation. If implementation of Directives could come about through voluntary or negotiated agreements member states might be more willing to agree to more ambitious targets. Industry may play a more constructive role if it has a greater say in implementation. Without the regulatory framework, preferably at EU level, such agreements will be impossible to monitor and are unlikely to deliver significant environmental change.

Market mechanisms and voluntary agreements are useful *additions* to the array of policy tools available for delivering sustainable development. They have not replaced command and control regulations, and are unlikely to do so. The UK government has said a great deal about new instruments but has, so far, not used them extensively. However, as pressure to deal with environmental issues inevitably increases, their application will proliferate. The key is to ensure that they are used in a proactive, not reactive, manner.

Acknowledgement

Helpful details on voluntary agreements were provided by Halina Ward of the Foundation of International Environmental Law and Development (FIELD).

References

Bates, R. and Morris, J. (1994) *Global Warming: Apocalypse or Hot Air?* (London: Institute for Economic Affairs).

Beckerman, W. (1975) *Pricing for Pollution*, Hobart Paper 66 (London: IEA).

British Petroleum (1995) *Economic Instruments for Environmental Protection* (London: BP).

Budget Statement (1994) *Budget Statement, 27th November 1994. Rt Hon. Kenneth Clarke MP* (London: House of Commons).

Confederation of British Industry (1994) *Environment Costs* (London: CBI).

Department of the Environment (1990) *This Common Inheritance* (London: HMSO).

—— (1995a) *Making Waste Work*, CM3040 (London: HMSO).

—— (1995b) *British Government Panel on Sustainable Development, First Report*, January (London: Department of the Environment).

—— (1996a) *Producer Responsibility for Waste* (London: HMSO).

—— (1996b) Speech by Rt Hon. John Gummer MP to the UN Framework Convention on Climate Change, 17 July 1996. Department of the Environment Press Release, London.

Department of Trade and Industry (1996) *Advisory Committee on Business and the Environment, Sixth Progress Report*, April (London: Department of Trade and Industry).

Department of Transport (1996) *Transport: The Way Forward* (London: HMSO).

Duncan, A. (1996) 'Developments in environmental regulation – implications for industry and regulators', *Journal of the Institute of Environmental Management* 3(4): 37–40.

ENDS (1977) 'Chemical industry deal on energy sets poor precedent', *ENDS Report*, November: 38–9 (London: Environmental Data Services).

Hewitt, P. (1989) *A Cleaner, Faster London: Road Pricing*, Transport Policy and the Environment (London: Institute for Public Policy Research).

HM Treasury (1997a) *Environmental Taxation: Statement of Intent*, HM Treasury Press Release 4, 2 July (London: HM Treasury).

—— (1997b) *Pre-budget Report*, November, CM 3804 (London: HMSO).

Hutton, S. and Hardman, G. (1993) *Assessing the Impact of VAT on Fuel on Low Income Households: Analysis of the Fuel Expenditure Data from 1991 Family Expenditure Survey*, (York: Social Policy Research Unit, University of York).

Jacobs, M. (1990) *Sustainable Development: Greening the Economy* (London: Fabian Society).

Jenkins, T. (1996) *A Superficial Attraction* (London: Friends of the Earth).

Johnson, P., McKay, S. and Smith, S. (1990) *The Distributional Consequences of Environmental Taxes* (London: Institute for Fiscal Studies).

Labour Party (1990) *An Earthly Chance* (London: Labour Party).

—— (1994) *In Trust for Tomorrow* (London: Labour Party).

NSCA (1993) *Pollution Handbook* (London: National Society for Clean Air).

Owens, S., Anderson, V. and Brunskill, I. (1990) *Green Taxes: A Budget Memorandum* (London: Institute for Public Policy Research).

Pearce, D., Markandya, A. and Barbier, E. (1989) *Blueprint for a Green Economy* (London: Earthscan).

Pearson, M. and Smith, S. (1990) *Taxation and the Environment: Some Initial Evidence* (London: Institute for Fiscal Studies).

Pigou, A. (1920) *The Economics of Welfare* (London: Macmillan).

RCEP (1972) *Royal Commission on Environmental Pollution. Third Report: Pollution in Some British Estuaries and Coastal Waters. Minority Report by Lord Zuckerman and Wilfred Beckerman*, CM5054 (London: HMSO).

Smith, S. (1995) *Green Taxes and Charges: Policy and Practice in Britain and Germany* (London: Institute for Fiscal Studies).

Taylor, R. (1992) *A Market in Environment* (London: Adam Smith Institute).

Tindale, S. and Holtham, G. (1996) *Green Tax Reform* (London: Institute for Public Policy Research).

Water Services Association (1993) *The National Metering Trials Working Group: Final Report* (London: WSA).

Wilkes, A. (1995) *Government Policies as the Catalyst for the British Environment Industry* (London: Environment Industries Commission).

3

NEW ENVIRONMENTAL POLICY INSTRUMENTS IN GERMANY

Wolfram Cremer and Andreas Fisahn

Our thesis is that the preference for certain instruments can only properly be understood in the context of the socio-political situation and its historical background. Discussion about instruments should be seen in the light of various conflicting interests. It makes little sense to hold the discussion about instruments in isolation as environmental protection must be placed in the context of the discourse regarding a new manner of regulation.

Command and control regulation instruments

The spirit of the 'economic-miracle country' initially did not permit debate about trivialities such as environmental protection. Politics was bound up in the magic of economic prosperity and technical innovation. In the 1950s the Water Resources Management Law (*Wasserhaushaltsgesetz* (WHG), of 27 July 1957) was enacted, which today still plays an important role in environmental law, having undergone various changes to increase the level of protection and develop certain instruments. At the time the WHG was enacted its primary objective was not environmental protection, but to ensure an adequate water supply for the population, which was in turn part of modern health provisions (Kloepfer 1994: 59, 84). Not only the WHG but also the water supply authorities viewed the supply of water as a public duty which required state control of resources. At the time of its enactment the WHG constituted an instrument typical of command and control regulation as enforced by the police. It regulated individual permits for water use according to the principles of administrative intervention (Gieseke 1992).[1]

At the end of the 1960s and in the 1970s environmental protection became a central topic of social discourse and debate in the Federal Republic of Germany (FRG). In the beginning the environmental policy discussion and legislation was motivated by anthropocentric, humanistic and social factors. It was about absorbing the consequences of accelerated industrialisation as they impacted on man's living conditions. It was about the humanisation of life and, in particular, socially weak classes who lived near industrial sites and were particularly affected by their emissions (Umweltprogramm der Bundesregierung 1971; Brandt 1961). The objectives and motives underlying environmental protection were emphasised in Willy Brandt's call for, and promise to create, blue skies over the 'Ruhr area'.

Like many other European states, Germany responded to growing environmental concern with state intervention and with command and control regulatory instruments. In the 1970s the German federal parliament enacted important environmental laws which remain in force today in substantially unchanged form. The following are the most important: the Waste Law of 7 June 1972 (*Abfallgesetz*); the Federal Law of Protection against Harmful Effects on the Environment of 15 March 1974 (*Bundesimmissionsschutzgesetz* (BImSchG)); the Amendment to the WHG of 26 April 1976, setting standards for permitted levels of emissions from the discharge and elimination of sewage; the Federal Conservation Law of 21 December 1976 (*Bundesnaturschutzgesetz* (BNatSchG)); and the Chemicals Law of 16 September 1980 (*Chemikaliengesetz*).[2]

The aforementioned laws are generally characterised as classical command and control regulation instruments, regulations based on the 'defences against danger' contained in nineteenth-century 'police and industrial law'.[3] The 'defences against danger' concept (*Gefahrenabwehr*) uses the 'requirement to obtain approval' instrument for certain activities which are potentially harmful to the environment. The granting of approval is dependent upon compliance with certain (more or less precise) legal requirements. The typical example is the need to comply with limits or technical standards (for example 'best available technology') when emitting certain substances into the air or water.

This command and control regulation dimension only covers part of the concept of reform of environmental legislation in the 1970s. It was embedded in a conception of global management, which (if only partly) was reflected in environmental legislation. Since the 1950s the model of the social-market economy was the hegemonic political conception. Despite different nuances within the political parties, it was undisputed that the social dimension of the market economy does not occur automatically, and is certainly not an inherent part of the market economy. Rather, the social dimension must be established by the state and protected against market forces. There was consensus across all political parties that state intervention was necessary in order to provide for people's basic existence and social welfare.[4] With the first signs of an economic recession at the end of the 1960s, the view that the state's social intervention should extend into political-economic intervention grew in popularity. The involvement of the Social Democrats in government (1966) led to the adoption of the

Keynesian instruments of economic global management and deficit spending as government policies (Ehrig 1989; Körner 1986).

The conception of politically induced global management of the economy was applied to other areas of society. Two such areas – the attempt to increase land planning (*Raum- und Landschaftsplanung*) and specific environmental planning (e.g. for water) – were of relevance for environmental protection.[5] Global management was supposed to be supplemented at a decentralised and operational level by the expansion of rights of participation,[6] which today are still contained in environmental legislation or other regulations (such as the Road Laws) which are used for environmental protection. The regulation of the relationship between man and nature was understood as part of this conception of global management. Even if environmental protection was not originally a central issue in this relationship, environmental protection groups slowly gained greater influence over what planning authorities considered to be the public good through rights of participation in the planning process (Steinberg 1993: 19).

The political programme of global management of the relationship between man and nature was never effectively realised by plans, which have to balance and take account of different local and regional interests. Its inadequate implementation is mainly due to the fact that the political programme of global management quickly came under pressure from two sides – on the one hand, from the environmental movement's detachment from, if not hatred of, the state and, on the other hand, from the growing influence of neo-classicism in economic policy. The growing importance of the latter was based on barriers on national economic management in the light of international financial integration.

Command and control regulation faces a dead end

The preference for any particular environmental law instrument does not stem only from the development of environmental problems, but is bound together with broader social and historical developments. Throughout the late 1970s and 1980s existing paradigms underlying the command and control approach came under increasing pressure. Four of the primary sources of pressure were an emerging ecological movement critical of technology; the limitations of classical economic theory, which privileged state intervention; the perceived environmental ineffectiveness of overregulation; and concern for maintaining German economic competitiveness in increasingly globalised markets (the 'location Germany' debate).

Criticism from the ecological movement

The basic direction of German environmental policies in the 1970s was technological. The traditional view that technological progress (namely the development of productivity) leads to economic and social prosperity for all dominated, and was applied as the solution to the 'recently recognised problems' of environmental protection. The majority view was that environmental problems could be solved by technological progress and environmentally friendly technology.[7] This view is also reflected in the regulation of limits on emissions and the orientation of environmental protection according to 'general rules of technology'. The technical orientation of policies became an issue of social debate within the topic of nuclear policy. The decision to use nuclear energy had been made long before. The conflict, however, first became virulent in connection with the political changes in 1968.

The comparatively widespread movement in Germany against nuclear energy was based on two political traditions: the pacifism and anti-military sentiments which developed after the Second World War, and the anti-capitalist sentiment of the student revolution. In the social debate about remilitarising Germany it was feared that creating a new army would lead to the possibility of that army being armed with nuclear weapons. With the planned expansion of nuclear energy, in particular by reprocessing atomic waste and by 'fast-breeders', the technical know-how and the actual ability to create nuclear weapons seemed to exist. Accordingly, the criticism of nuclear energy was interwoven with the criticism of atomic weapons, which was an important factor in the development in Germany (unlike most other European countries) of a wide public movement against nuclear energy. There was great sensitivity towards warnings about the use of nuclear fission, and this turned into public protests.

The ecological movement did not stop at protests against nuclear energy. Nuclear energy became the paradigm for the risks and dangers of developing technology and 'progressing industrialisation'. Characteristic features of the ecological movement were the criticism of Western and Eastern industrialisation, of economic growth, and the distant relationship between new technology and the planning and administering state.[8]

Part of the debate about nuclear energy took the form of direct physical confrontations between opponents of nuclear technology and the state. Not least, this conflict with the 'social-democratic' state led to deep-seated scepticism in the ecological movement about the possibility of sustainable state regulation of the relationship between man and nature. The state seemed to be too closely interwoven with economic interests and too focused on technological solutions to support environmental protection actively. The environmental movement preferred decentralised forms of organisation, which could supposedly bring about a different economy and a different relationship with nature. The relationship with state institutions was characterised more by the classic liberal defence against restriction of freedoms and the imposition of ready-made decisions than by the possibility of state regulation of the environment.[9] It is thus not surprising that the previous consensus in favour of state intervention and command and control regulations came under pressure from the environmental movement of the Left.

Shifting economic paradigms

At the same time, other trends also influenced the end of the state interventionist paradigm. The hegemony of neo-classical or monetarist economic theory was of socio-political importance. The classic view that politics as the centre of society could intentionally and effectively manage and regulate the relationship between man and nature was increasingly challenged by social scientists, particularly Luhmann (1984; 1986: 220), and was also important for changing the paradigms relating to environmental law regulation.

In addition to the perceived inability of state intervention to manage complex societies, traditional economic theory came under further pressure in the mid-1970s as the international framework conditions of the economy changed. National demand-oriented economic policies were stretched to their limits and were not able to solve the problems. After 1973 there were deep-seated changes. As a result of the oil crisis production costs increased considerably and the economies of oil-importing states went into recession. The system of fixed exchange rates (Bretton Wood) was abandoned. As a result, the leading currency, the dollar, came under continuous pressure to devalue and in the 1970s had to accept a decrease in value of an average of 17% compared to other industrial countries (Altvater 1992: 149). This resulted in more imports and fewer exports for demand-oriented anti-cyclical economic policies in other states. At the same time the rate of inflation rose and, with it, the demand for credit. At the end of the 1970s increased demand for credit and high-interest policies in the US resulted in extremely high interest rates worldwide (Hübner and Stanger 1986: 69). This development was seen as a loss of national sovereignty over interest rates and brought Keynesian economic policies even more into question.

Based on the change of paradigms in US economic theory (for many of these, see Friedman 1962), the influence of neo-classical or monetarist economic policies grew in Germany. These questioned the chances of success of demand-oriented and anti-cyclical global management in economic policies, and demanded deregulation and less state intervention in the economy. It was said that state intervention, by imposing restrictions on business activity and engineering socio-political redistribution, damaged the balance of economic development, reduced profits, and led to less investment and slower growth (Herr 1988: 66; Kalmbach and Kurz 1983: 57).

The proper form of state intervention also became an important aspect of the shifting economic paradigms. Discussion about economic and market-based instruments of environmental policies initially took place among economists and became influential as the 'new political economy' (Kirsch 1993; Horbach 1992). Based on utilitarian concepts, the central theory was put forward that actions could be more effectively managed by systems of incentives than by prohibition and punishment. It was thus demanded that there be an environmental political change from command and control regulation to systems of market-force incentives, which was widely welcomed within the environmental law discussion and by environmental groups as their confidence in the effectiveness of regulation of command and control regulation had been badly shattered long before.[10]

While it might have been expected to pose a formidable obstacle to any regulatory reform which did not fundamentally alter the trajectory of social development, and which encouraged greater reliance on market mechanisms, even the Green Party was amenable to the new environmental instruments. The changes within the Green Party are exemplary of the shift of powers within the political arena in the 1980s. At the beginning of the 1980s the Green Party was split into two approximately equal-sized factions: the pragmatic politicians (*Realos*), who wanted to follow policies of gradual reform in alliance with the Social Democrats (SPD), and the fundamentalists (*Fundis*), who totally rejected the notion of a coalition with the 'established parties' and considered fundamental changes of the system to be necessary. At the end of the decade the 'fundamentalist' positions had been pushed out of the Green Party and their leading figures had either left, modified their political programme or become insignificant.

Ineffective regulation

Scepticism about state regulation was also expressed through criticism of 'overregulation' (Voigt 1980) and of the environmental effectiveness of command and control regulation. Overregulation was (and is) seen as a dominant trait of state intervention and must be understood as a tendency to regulate all areas of society's life by legal rules. Concerns were initially expressed about the 'flood of regulation' (Vogel 1979: 321) as it infringed upon freedom. The growing number and complexity of regulations was considered a factor in undermining their effectiveness, as people are willing to follow the law only if they are aware of it. This could no longer be expected with more comprehensive and more detailed regulation.

Legal rules in environmental law were increasingly viewed as highly ineffective. A flagrant lack of enforcement of environmental regulations was apparent (Mayntz 1978). Today, environmental law discussions generally view lack of enforcement as a central problem (Wagner 1995: 1046; Lübbe-Wolff 1993: 217; Lahl 1993: 249). This results in the empirical fact that the environmental-protection objectives of the law are not optimally realised. The regulations are used as tools for discussion and as a means of exerting pressure in a 'bargaining process' between officials and site operators, but not as binding regulations. This process of negotiating environmental standards at the lowest level of law enforcement has been characterised as an 'informal legal state' (Bohne 1981; Winter 1985: 210; Hennecke 1991: 267).

Economic competitiveness and the 'location Germany' debate

With the opening of the borders in the East and the accession of the German Democratic Republic (GDR) to the FRG in 1990, the political discourse moved in the direction of an environmental-political 'roll-back'. Whilst environmental protection has not actually been abandoned as a political objective, it is increasingly simply paid lip-service and merely referred to symbolically. The most striking example of this is ozone legislation, where, against the advice of the environmental ministry, the levels for banning driving in the event of summer smog have been set so high that the laws are, in effect, irrelevant. The demand to 'free' the economy of social and environmental restrictions in order to render 'location Germany' more attractive was considered of greater importance.

It is predicted that the whole economy will become internationalised,[11] which is considered an unavoidable consequence of the advancements in communications and the 'liberalisation' of world markets. Fears have been expressed that a location problem would arise for Germany from this internationalisation of markets. The result of these processes would be even greater competition, no longer predominantly at a national level but increasingly at an international level. As the individual nation-state can not counter this competition (for a different view, see Göll and Schuster 1995: 99), industry, in particular, calls for an improvement in the national conditions in Germany in light of international competition. 'Location Germany' should be made more attractive and the national economy should be prepared for competition with other national economies (Hoffmann 1988: 148; Klodt and Stehn 1994). To this end, in particular, costs of production should be decreased and investment facilitated. Both of these objectives would be achieved by, amongst other measures, deregulation in environmental law and by removing obstacles to investment imposed by environmental law.

German unification focused particular attention on improving the investment climate in the former GDR. After the new federal German states had been almost completely de-industrialised by the introduction of market competition, the new official objective was to 'catch up' in respect of their industrialisation. In particular, there is supposed to be an accelerated improvement of the infrastructure, and private investments are supposed to be capable of being realised at an 'accelerated' pace. Many have argued that restricting economic activity on environmental grounds hinders this attempt to catch up in respect of industrialisation.

Environmental groups and environmental politicians (one can hardly still speak of an ecological movement) could hardly counter this discourse.

Having introduced the cooperative and market conforming instruments, the logic of the 'location discourse' seemed plausible. The national competitive situation was accepted as the premise for cost-neutral ecological tax reform. Consumption of resources and the discharge of harmful substances should be made more expensive by taxation, employment of labour cheaper. It was hoped that a burst of innovation in environmental technology and positive employment effects would be achieved by means of this reform (Draeger and Wolf 1995: 1071; Ewringmann 1994: 43). However, as is discussed in more detail below, the governing coalition showed no inclination to adopt these concepts in any form. It was argued that the competitive situation of the German economy would become even worse as a result of comprehensive tax reform. Even among advocates of ecological tax reform there is little consensus on many such questions, to the extent that one can hardly speak of a uniform and consistent overall concept.

The search for new instruments

Having demonstrated that the environmental law discourse can not be understood outside its overall social context, this section goes into more detail about the use of new environmental instruments in Germany, and tries to explain why these instruments have been only slowly legally implemented to date. Environmental taxation is comprehensively described because these instruments are at the centre of the discussion in Germany.

The deficiencies of environmental political management by command and control led to a search for alternatives.[12] This search led to different instruments of environmental law management. In summary, the following categories of environmental law instruments exist:[13]

1 Instruments which seek to improve command and control regulation, distinguishing between:

- the improvement of control and supervision by officials, the public or independent institutions;
- increased flexibility.

2 Procedural improvement and cooperation, distinguishing between:

- cooperation through rights of participation for individuals, the general public and environmental groups;
- the participation of experts and informal regulations (cooperation between the state and the economy).

3 Economic management:

- direct financial incentives through comprehensive ecological tax reform or at least environmental taxes in specific areas; certificate and compensation models; and invitations of tenders for public contracts;[14]
- individual liability for environmental damage; liability funds for environmental damage and insurance models;
- indirect market conforming incentives (labelling; ecoaudit).

It is apparent in the environmental law discourse that preferences have moved towards economic management (Beschluß der Bundesregierung 1994: 154, 161). This means that the discussion of command and control regulations which was critical of the state and the market economy has lost ground in favour of market-based solutions.[15]

Environmental taxation

Command and control law separates legal and illegal environmental uses and provides no economic incentive for greater compliance with the required standards than is legally necessary. In contrast, environmental taxes are aimed at imposing costs on every environmental use and at creating incentives to minimise environmental uses within the command and control laws and prohibitions (see Chapter 1). The imposition of environmental taxes is justified on the basis of the polluter-pays principle (Hendler 1990: 577; Breuer 1992: 485; Köck 1993: 59; Endres 1994: 13). For example, water and air for a long time qualified as free natural resources. It did not cost anything to use these goods (apart from the precautionary efforts to comply with the emission levels), although their use (even in compliance with the levels) resulted in external costs (so-called negative external effects). Health costs from environmental pollution, increased costs of preparing drinking water, and the costs of treating sulphur dioxide damage to forests and buildings should be mentioned. The form of market failure (Endres 1994: 13; Behrens 1986: 86)[16] considered here, namely the creation of external costs (to be distinguished from external uses), results in the produced goods or services being relatively 'too' cheap for their producer. These costs are borne partly by the general public and partly by unconnected individual third parties.[17]

In the reform discussion many instruments are mentioned to eliminate the aforementioned external effects, by having the polluter internalise the costs (Endres *et al.* 1993; Klocke 1995; Keppler 1994: 121). Apart from liability law and certificate models, each of which is discussed below, taxation of environmental uses has received particular attention.[18] Environmental economists suggest that, in accordance with a model developed by Arthur Cecil Pigou, in order to eliminate negative external effects the producers of such effects should pay a tax corresponding to the marginal costs arising in a socially optimal situation (Pigou 1952: 191).[19] From an economic perspective, it provides the principles for politically more acceptable variations of taxing the external effects which could supplement command and control regulation. The (partial) internalisation of costs resulting from environmental uses, achieved with the help of environmental taxes, aims to motivate enterprises and individuals to develop environmentally friendlier avoidance strategies. In view of the allocation effects, the taxation solution is, at least in theory, a superior model to the command and control law, because it facilitates an economically more efficient use of environmental goods (Hansmeyer and Schneider 1992: 12; Wicke 1991: 135). Extensive use has been made of environmental taxation in German environmental policy. Proposals include federal taxes on sewage and traffic, as well as a carbon-dioxide tax or an energy tax, sulphur-dioxide taxes, packaging taxes, local transportation taxes, road-use charges, chloride taxes and ground-water taxes. In view of the costs and deficiencies of enforcement, taxes might be preferable to command and control. The enforcement-friendly potential of taxes is the capacity to activate the interests of taxpayers to avoid pollution (Lübbe-Wolff 1996b: 103; for views which dispute this, see Dose 1990: 377).

Sewage tax

Unfortunately, the Sewage Tax Law immediately repudiates the idea of innovation through economic instruments. It was enacted in its original form on 13 September 1976, at the same time as the most important command and control regulation environmental laws. Pursuant to the Sewage Tax Law, dischargers of sewage are required to pay tax, the amount of which is determined according to the level of harmfulness and the amount of sewage. The goal of this instrument was primarily to initialise a shift in behaviour, to go below the limits.

In 1990 M. Jass presented a study of the effectiveness of the Sewage Tax Law, the central conclusions of which were that over the years the water law requirements led to considerable improvements of the purification of sewage and the environmental technology of the companies studied, but that the requirements of the command and control regulation were the central instrument. The sewage taxes led at most to accelerated implementation of these requirements, but did not constitute independent incentives for further improvements (Jass 1990).

The limited effectiveness of the sewage taxes can, however, be put down to a combination of less pollution and falling taxation. Because of the relatively high taxation imposed on pollution which is prohibited pursuant to the command and control regulation, sewage taxes only create an incentive to comply with the command and control regulation, but nothing more (Lübbe-Wolff 1996b: 112).

Petrol and vehicle tax

In order to counter air pollution from vehicles, drastically increased petrol prices and increased taxation have long been discussed as a means to reduce the use of private cars. However, taxation of petrol and diesel still primarily constitutes a source of finance for the state; the control of behaviour is at most a secondary effect. This is evident from the considerably lower taxation of diesel, which essentially constitutes preferential treatment for freight traffic.

An Act of 22 May 1985 established a tax advantage for vehicles with catalytic converters, and a further drastic increase in the vehicle taxation of cars without catalytic converters was introduced in 1997. As a result of this advantage and the tax advantages of lead-free petrol, new vehicles are only available with regulated catalytic converters. From an environmental perspective, these tax advantages have, prima facie, only had a positive effect. In the short term, however, the instrument is double-edged, as it has simultaneously created an incentive to buy new cars (leading to ecological damage as a result of production and scrapping).

Environmental taxes at local and state level

Economic taxation was also introduced at local and state level. Various states have imposed the following: requirements to pay compensation for using conservation areas and woods, and for extracting water (the so-called *Wasserpfennig*); waste taxes; and, in North Rhine-Westphalia, licence fees for disposing of harmful waste. On the local level, taxes on 'one-way waste' (packaging in which food and drinks are sold and which can only be used once is burdened with a tax) have been imposed by certain municipalities.[20] As far as one can tell from experiences with the one-way waste taxes, they are efficient. Most of the manufacturers do not use one-way packaging any longer.

Carbon/energy tax

Despite having been approved of, in particular by economists (Keppler 1994: 121; Endres 1994: 177; Cansier 1994: 642), the implementation of the 'environmental taxes' instrument has not experienced uniform success in Germany. In particular, the energy or carbon-dioxide tax (or a combination of taxation of energy use and carbon-dioxide emissions, a carbon-dioxide/energy tax), which has long been vehemently discussed among the public and within political bodies, is awaiting its legal implementation.

Some economists have expressed a belief that the tendency of German lawyers to cling to the command and control regulation is partly the cause of this delay (*ibid.*).

Explicitly formulated resistance came, in particular, from German industry, although it did not present itself as a monolithic block here. The failure to adopt a carbon-dioxide/energy tax is due in particular to the obligation to reduce carbon-dioxide emissions and specific energy consumption by the year 2005 by 20 per cent of 1990 levels (X 1996: 3; Bundesverband der deutschen Industrie 1996: 3).

For a long time the views expressed by the representatives of trade unions were not unified. After years of internal debate, on 4 June 1996 the federal board of the Association of German Trade Unions (*Deutscher Gewerkschaftsbund*, DGB) called unanimously for the introduction of an energy tax, which the chemical, paper and ceramic trade unions, in particular, had long rejected.[21] According to this resolution, the energy tax should be imposed on all fossil fuels and electricity (except electricity from regenerated sources). Tax reductions and exceptions are envisaged for enterprises which are heavily energy-dependent and energy-intensive, which ultimately might have facilitated the approval of chemical, paper and ceramic trade unions.[22]

The implementation of a carbon-dioxide/energy tax in Germany cannot be expected imminently. After years of controversial discussions and changing statements, the federal government (led by the Christian Democrats (CDU; CSU in Bavaria) and the Liberal Party (FDP)) decided not to follow an isolated national path, at least at the moment. It is currently seeking a solution at European Union level (Beschluß der Bundesregierung 1994: 161).

Important members of the Social Democratic Party (SPD) have also spoken out against a national carbon-dioxide/energy tax in view of recent dramatic changes in the German employment market.[23] However, on the whole the party favours the introduction of such a tax in Germany if a carbon-dioxide/energy tax cannot be implemented at European Union level (SPD 1995: 127, 129, 132, 142).

Only the Green Party has long been in favour of Germany following the path of adopting a carbon-dioxide/energy tax unilaterally, given the difficulties in implementation at the European Union level. Within the party, the question as to how the funds obtained from the tax would be used was long disputed. One section of the party wanted to use the funds for environmental protection; other sections to reduce employment costs. The latter has now become the prevailing view within the party, even if it is felt that part of the funds should be used for ecological investment programmes (Länderrat von Bündnis 90/Die Grünen 1995: 1, 6).

The environmental-political management effect of environmental taxes is, to a large extent, dependent upon the actual amount of the tax. Only a sufficiently high tax is likely to influence the behaviour of the affected entity significantly. In view of this, the above-mentioned sewage tax is ineffective from an environmental management perspective, and it is not surprising that the carbon/energy tax encountered determined resistance. Although from an economic perspective environmental taxes correspond to the polluter-pays principle, which is an inherent element of the market economy, their politically stipulated objectives can engender as much resistance as other instruments of environmental regulation.

The rules exempting energy- and/or emission-intensive industries which were widely demanded as a result of international competition should be judged very sceptically. One should criticise the fact that, as a consequence of these exemptions, quantitatively relevant environmental uses are not covered and industries which cause a particularly large amount of environmental pollution are spared. In addition, necessary structural changes are blocked or steered in the wrong direction. This is because the products of the industries which enjoy the exemptions[24] not only may be in international competition, but also may be in (potential) substitute competition with environmentally friendlier products from enterprises which do not profit from the exemptions, be they domestic or international. In determining the exemptions these interdependent factors should be taken into account, so as not to improve the terms of competition for energy- and emission-intensive goods.

In view of the costs and deficiencies of enforcement, an energy tax would possibly be preferable to an emission tax because the latter would require greater efforts as regards obtaining information and monitoring. However, liberalisation of the energy market on the EC level will probably increase the enforcement costs of an energy tax (Keppler 1994: 129).

Principle of Cooperation and Waste Law

An example of 'gentle' economic management by cooperation is the introduction of the 'Dual System' in connection with waste disposal. In order to reduce packaging waste, at the end of the 1980s the federal government planned to introduce an obligation for shopkeepers to take back sales packaging. It was hoped that this would indirectly influence behaviour, in that shopkeepers would convert to offering products with little or no packaging, which would eventually do away with packaging and waste even at the production stage. This model was legally entrenched in the Packaging Ordinance of 12 June 1991. However, after prolonged negotiation with trade and industry the ordinance was supplemented with an alternative option, a system of separation of waste. In practice, it is this so-called Dual System of Germany (*Duales System Deutschland* (DSD)) which has been established.

The system functions as follows: the producers pay a fee to a private waste disposal company, the DSD. In return they have the right to include a 'green point' logo on their products' packaging. The consumers are required to collect together all packaging waste containing a 'green point' logo and place it separately in a special yellow rubbish bag, which will be collected at regular intervals by the waste disposal company. The consequences of this cooperation model can be outlined as follows. First, instead of avoiding waste, as was initially intended, there is merely a separation of different types of waste. So far there have not been considerable reductions in packaging. In its five-year financial statement the DSD declared that a reduction in packaging waste had been achieved. Packaging waste had dropped by 900,000 tonnes in the five years. The weight is, however, of only small importance. Environmental groups justifiably point out that the packaging industry itself reports continual production growth, as a result of which a drastic reduction of packaging waste cannot be a realistic expectation (*Frankfurter Rundschau*, 25 June 1996: 13). Second, the consumer bears the burden in that he finances the profits of the waste disposal companies by way of increased prices.[25] In addition he continues to pay waste charges to the public waste disposal authorities. Ironically, it has not been possible to decrease the charges correspondingly despite smaller amounts of waste because it remains necessary to maintain the waste disposal plants and their infrastructure. Third, packaging waste is partly recycled. Although this is a positive factor, the amount of waste recycled at the moment remains below the desired levels.[26]

Certificate models

Certificate models, which are viewed positively in the economic literature, are given less consideration in public discussion (Endres 1994: 165; Keppler 1994: 121). The core of these models is the idea that political decision-makers will stipulate an overall maximum emission level (or overall energy use) for a certain harmful substance in a certain region and will allocate the rights to use this 'environmental capacity' amongst the environmental users. Basically, two models are being discussed to allocate the rights, namely auctioning and politically stipulated allocation. According to the latter variation, the allocation should be based upon the amount of harmful substances emitted by the enterprises within a certain period of time.

These rights will then be documented in 'emission certificates' and will be tradeable (Endres 1994: 106; Endres *et al.* 1993); only to this extent do the certificate models contain a market-economy element and follow the objective of efficient allocation. The certificate model has been applied in certain sectors, in particular in the US (Endres 1994: 113). The basic idea of the certificate concept was adopted in German law by the amendment to the BImSchG and formalised in the Technical Direction to Keep the Air Pure (TA Luft) in 1986. The compensation model allows emissions to exceed certain limits in one site if the operator of the site or third parties overcompensate with reduced emissions in other sites. The compensation possibilities (Rohde 1990: 159) thus achieved were slightly improved by a new amendment to the BImSchG in 1990 (Goßler 1990: 255; Schlabach 1990: 251). However, little use has been made of the instrument (Gawel and Ewringmann 1994: 121; Endres 1994: 116), probably because compensation is required within a small region in one calendar year and, according to the TA Luft, exemption from meeting the limits was granted only for a fixed time period in the past (Gawel and Ewringmann 1994: 120). The amendment to the BImSchG in 1990 has still to be concretised by an amendment to the TA Luft which might offer compensation permits for longer periods.

The environmental-political value of certificate models depends (as with environmental taxes) substantially on the politically stated aims (from an economic perspective an exogenous factor). That is, the stipulation of the overall permitted amounts of emission in a region is decisive. As with environmental taxation, failure to implement these instruments is due to their management implications.

Particularly in the introductory phase, additional structural problems exist which render doubtful the success of the certificate model as a management instrument to reduce emissions in individual enterprises, at least in the short term. The demolition of existing and approved old sites will hardly be worthwhile for financial reasons.[27] As far as enforceability is concerned, the certificate model has no advantages over the command and control regulation as its administration (even if done by private bodies), as well as the continued need to monitor discharges of emissions, is relatively costly (Lübbe-Wolff 1996b: 134).[28] From an enforcement perspective, the allocation of 'energy use rights' would probably be more advantageous.

Liability for environmental damage and fund models

The Law on Environmental Liability (*Gesetz über die Umwelthaftung* (UmweltHG)) of 10 December 1990 introduced no-fault liability for most sites as of 1 January 1991. This was established to optimise protection of the environment and the legal position of persons who suffer damages by private law (Breuer 1995: 491). This instrument differs from command and control as it tries to internalise negative external effects. The law is applicable to environmental damages to air, water and soil caused by substances, noise emissions, radiation, gases, steam and heat. Despite resistance from industry, the liability covered not only actual harmful incidents but also damage from permitted regular operations (Bender *et al.* 1995: 54). Only in the event of *force majeure* does the liability not apply. Furthermore, the UmweltHG renders proof of causation easier, but does not suffice to attribute actual distance and summation damages to individual entities as it is not possible to fulfil the (constitutionally) necessary minimum requirements to prove causation. This is also the reason why the UmweltHG was hardly relevant in practice.

In our view, this indicates the central problem of the UmweltHG as well as generally of any environmental liability law geared towards individual polluters. This problem should not be dealt with by the state having to accept liability,[29] but by means of a supplemental fund. Although the rules would be difficult to draft, it would also be conceivable for the fund to have a right of recourse against the individual polluters (Rehbinder 1992: 120). Only the fund solution corresponds to the polluter-pays principle (Kinkel 1989: 297; Rehbinder 1992: 120) (to this extent to be understood collectively) and the internalisation model. From a management perspective such a model would be welcomed, since, besides individual liability, it would increase the avoidance incentives because of the additional risks (although given that it is collective liability the influence may not be significant). There would be a further directing effect if the amount of the contributions payable by the individual enterprises was determined not according to business-management factors but according to the level of emissions. This variation of the model, which is determined in accordance with the individual polluter-pays principle, is thus preferable despite the increased enforcement costs and the likely enforcement deficiencies (Kinkel 1989: 298).

The 'Acceleration Amendment'

As mentioned earlier, in recent years the dominant issue in environmental legislation has been 'Acceleration'. Rather than introducing new instruments, this has sometimes caused an actual reduction in environmental law procedural standards. The 'Acceleration Legislation' was initially justified on the basis of the need for quick construction measures in the five new German states and began with the Road Construction Acceleration Law (*Verkehrswegebeschleunigungsgesetz*) of 1991, which was specifically for these new states. Moreover, the Investment Measures Law (*Investitionsmaßnahmegesetz*)[30] specifically provided for certain traffic projects in the east of Germany so that these projects were not required to follow the usual administrative procedure. Having 'tested' it in eastern Germany, the federal government wanted to apply the accelerated road-construction planning system (with the exception of certain details) to the west of Germany (Steinberg 1993). This was done in 1993 with a law to simplify the planning procedure for roads (*Gesetz zur Vereinfachung der Planungsverfahren für Verkehrswege*). At the same time the Expansion of Highways Law (*Fernstraßenausbaugesetz*) and the Expansion of Railways Law (*Bundesschienenwegeausbaugesetz*) were passed, legally stipulating the need for these transportation routes.

The Law to Facilitate Investment and Residential Building

(*Investitionserleichterungs- und Wohnbaulandgesetz*) transferred substantial procedural elements which had been given an accelerated effect to other areas of law (for example waste law and construction law). Furthermore, the new accelerated procedures apply in these areas for the whole of the Federal Republic of Germany. Moreover, in the new German states since the changes in the Federal Land Planning Law (*Raumordnungsgesetz*) land planning procedures are no longer required (*Raumordnungsverfahren*) if they would disproportionately delay investments. The requirement that the *Raumordnungsverfahren* be open to the public was replaced with a power granted to the states to regulate whether and to what extent the public should be involved.[31]

The erosion of rights of participation and approval requirements

An essential element of acceleration has been the erosion of rights of participation. The participation of individuals, the general public and groups in the transportation-route approval procedures and various other approval procedures (for example waste sites) was reduced by restricting the right to object and by partially cancelling the open hearing dates (Fisahn 1996: 63). This erosion of individuals' participation and consequently also of the involvement of environmental interests occurred, for example, through replacement of the *Planfeststellungsverfahren* by *Plangenehmigungen*. This led to the abandonment of environmental impact assessments (EIAs). Irrespective of the site's effect on the environment, it is not required that the *Plangenehmigung* include these assessments.[32] In addition, accelerated effects were sought by reducing the prescribed channels of appeal.

The 'Acceleration Amendment' led not only to an erosion of rights of participation, but also to the partial abandonment of the requirement to obtain approval. In addition, short cuts in examining the environmental needs and more environmentally friendly alternatives led to far less consideration being given to environmental interests in the administrative proceedings.

Germany does not appear to have stopped adopting acceleration legislation. Various commissions, working parties and ministries have produced and proposed more comprehensive 'acceleration catalogues' (Lübbe-Wolff 1995: 57). In particular, the proposals of the so-called Ludewig Commission and the Schlichter Commission's catalogue based on them should be mentioned. The Federal Environmental Ministry and the federal government have already adopted some of the suggestions of the Schlichter Commission and included them in legislative procedures.[33] Accordingly, for example, pursuant to the proposed amendment to sections of the BImSchG, in order to make certain changes to plants which require approval under current legislation the operator would in future simply need to submit a notice which substitutes the need for approval. It is intended that the proposed amendment will reduce the number of changes requiring approval. However, this procedure does not offer the same level of protection as results from receiving approval because the operator would run a greater risk of the authorities subsequently requiring him to implement further compliance measures. The fact that about 80 per cent of the procedures of the requirement to obtain approval pursuant to the BImSchG relate to changes to sites and only 20 per cent apply to new sites shows the quantitative relevance of this amendment.

Unlike many of the other instruments described in this chapter, acceleration measures are not (even as a side effect) based on any environmental-political motives. They appear solely as instruments to make the 'location Germany' more attractive,[34] and not as instruments of environmental-political management. It is nevertheless necessary to question the actual acceleration and simplifying potential of these measures, as well as the dangers for the environment associated with these measures.

The outlined erosion of the participation of individuals should be criticised from a legal-political perspective because of its supposedly negative consequences for environmental protection. Based on the philosophy of the European Court of Justice,[35] in European Community law the individual and environmental groups are attributed the role of protector of environmental interests.[36] This line of argument, developed by the Court for court proceedings in particular, can also be applied to administrative proceedings. The role of the individual and environmental groups as protectors of environmental interests is important because of limited administrative resources, in particular for monitoring environmental law. In addition, in the area of preventive control of environmentally relevant projects the authorities often do not possess sufficient information about the environmental interests affected. Public participation (at least in theory) achieves a more balanced representation of conflicting economic and ecological interests. In view of the objective to accelerate proceedings by reducing rights of participation, the first specific evaluations paint a relatively optimistic picture (in respect of the acceleration measures relating to motorway planning, see Paetow 1996: 57). It is, however, doubtful that the intended acceleration effects will play a significant role in making 'location Germany' attractive. Recent studies show that the length of approval proceedings plays only a minor role in investment decisions (Steinberg 1995: 209, n.4).

In respect of the partial substitution of the approval procedure by notice proceedings, even the efficacy of the intended acceleration effect is doubtful. The proposed amendments would cause an increase in the number of proceedings, which would complicate the activities of environmental administration. The administration must examine whether a simplified procedure is permissible in individual cases, which is inconsistent with the acceleration effect (Koch 1996: 220). It further complicates the officials' examination of which procedure to select and renders it relatively time-consuming for the officials to apply uncertain legal concepts, which particularly hinders enforcement (Mayntz 1978: 33, 349, 410; Lübbe-Wolff 1996b: 92; Koch 1996: 216). In addition, from an operator's perspective, the planned amendment does not appear particularly attractive because of the risk of retrospective orders. A representative of the Lower Saxony Business Association (Unternehmensverbände Niedersachsen e.V.), Wolfgang Rohde, agreed – in his view it was incomprehensible that industry had been greatly in favour of the proposed amendment to the BImSchG. It is possible that his prognosis that this instrument will hardly be used in practice will be proved correct. From an environmental-political viewpoint the notice procedure is problematic where construction is carried out following submission of notice and the operation of the site then conflicts with environmental law, because it is very difficult to enforce any retrospective orders.

In this context the point should also be made that when a procedural law is changed the related new demands upon the administration cause short- to medium-term procedural delays, simply because of the time necessary to adjust and the difficulty of applying uncertain legal concepts (frequently these are first implemented by way of an explanatory declaration by one of the highest levels of courts). Thus the removal and substitution of specific instruments and procedural models should not take place before an evaluation, which is often not the case (Paetow 1996: 60).

The introduction of the so-called 'star proceedings' should be welcomed without any reservations. These envisage all officials involved in the environmental procedure handling the matter at the same time. This procedure, which does not infringe material environmental law, is preferential to the previous practice, pursuant to which officials were involved with the procedure one after another, because of its acceleration effect. A further possibility for acceleration, which could also reduce lack of enforcement, would be to increase personnel within the environmental administration, but for cost reasons this is not feasible (Lübbe-Wolff 1996b: 10).

Finally, it should be noted that the German legislature was much more energetic in implementing the 'environmentally detrimental location policies' (Bender *et al.* 1995: 8, n.25) than the aforementioned environmental-protection-oriented instruments. The president of the Bundesumweltamt (Federal Environmental Protection Agency), Heinrich von Lersner, refers in this context to an 'ecological counter-reformation' (quoted in *ibid.*).

(Poor) implementation of EC law in Germany

As well as nationally initiated measures, the implementation of EU law has shaped national environmental law in recent years. While EC law imposed tools on all member states, it is useful to show the poor implementation of important environmental Directives in Germany. The outlined implementation of EC law in Germany should give a more comprehensive idea of environmental policy in Germany within the last decade.

In the past, numerous proceedings were instigated against Germany, which likes to claim a leading role in environmental protection (Everling 1992: 379; Breuer 1990: 80). These actions were based on incorrect implementation of EU environmental Directives and frequently resulted in judgements against Germany (Gellermann: 1994: 117, n.3). In respect of proceedings relating to environmental matters between 1988 and 1992, Germany ranked in the middle of a table produced by Christoph Demmke based on the *Tenth Report on the Application of Community Law* (COM(93) 320, 28 April 1993; Demmke 1994: 42). The EU Directive on the Freedom of Access to Information in the Environment (the national implementation of which is outlined and assessed below) was also first implemented by Germany after long delays.

Freedom of access to information on the environment

Germany delayed by almost one and a half years the implementation of the 1990 EU Directive on the Freedom of Access to Information on the Environment.[37] The German Freedom of Access to Information on the Environment Implementation Law (*Umweltinformationsgesetz*), however, in no way ended the discussion about the extent of rights to environmental information. The often ambiguous formulations of the German legislature and also the restrictive – and, from the perspective of European law, questionable – provisions of the *Umweltinformationsgesetz* contributed to this.[38] These uncertainties led to an overwhelming 'flood of publications' in Germany within a short space of time. By June 1996 over seventy articles in periodicals (Theurer 1996: 327) and four commentaries had appeared (Fluck and Theurer 1996; Schomerus *et al.* 1995; Röger 1995; Turiaux 1996).

The analyses and assessments of the many disputes would fill a book on this subject. We will only try to compare the principles of *Umweltinformationsgesetz* with the culture of German administrative law. Undoubtedly, the codification of rights to environmental information results in an important change in the general public's function in respect of a limited area of German administrative law. Whilst the laws of other states, including members of the EU, already recognised rights to information from the administration before the *Umweltinformationsgesetz* was passed (Engel 1993: 146; Bidner 1994: 408), official documents remain confidential in Germany, almost without exception.[39] The right to information conceptually expands the possibilities for the public and environmental groups to support the authorities in monitoring substantive environmental law, but also to control it. In view of the importance of the right to information for officials, industry and the environment, there are deep-seated differences of opinion in assessing this new environmental instrument. These range from euphoria[40] to scepticism (Erichsen 1992: 418; Schmidt-Aßmann 1993: 933) and, in part, to tastelessly discrediting[41] this instrument. The authors favour the positive assessments of this instrument because comprehensive information helps optimise the use of other environmental instruments.

The internalisation concept and command and control regulation are dependent upon knowledge of the extent of and the effects of the use of the environment (see Heyvaert 1997). An informed public can thus provide important assistance. Besides influencing environmental policy, the public can also follow enforcement informatively and, on the other hand, control the monitoring authorities. Ultimately, being informed and achieving transparency are conditions of the free-market economies within Germany and the EU, assuming one accepts that the environmental impact of a product is a parameter of competition. However, there is little cause for rejoicing, first, according to a survey performed by the newspaper *Öko-Test* in the summer of 1995, which found the environmental administration's knowledge and application of EIA unsatisfactory, and, second, due to a relatively restrictive interpretation of the *Umweltinformationsgesetz* by the German administration courts (VG 1995).

Ecoauditing

In December 1995 the Ecoauditing Law (*Umweltauditgesetz*) and a series of supplementary ordinances came into force in Germany to implement the EC Ecoauditing Regulation (see Chapter 10). Whilst the Regulation did not require implementation in the narrow sense of the word because of its direct effect pursuant to paragraph 2 of Article 189 of the EC treaty, in order to function the auditing system nevertheless required national laws about the admissibility of environmental experts and organisations of environmental experts, the supervision of reporting functions and the registration of examined locations of enterprises, and the appointment of the various responsible bodies.[42] Whereas the enactment of the German Ecoauditing Law was in principle supported by a wide political and social consensus (Lütkes 1996: 231), there was disagreement regarding important details between the interest groups involved in the legislative process.[43] As we do not have the space to analyse the provisions of the Ecoauditing Law in detail here, we shall only outline its overall tendency to favour industry, which could jeopardise the success of the auditing system.

During the legislative procedure industry and environmental groups disagreed about who should be entrusted with the task of determining the eligibility of environmental experts. Whereas environmental groups called for an 'authority model' and the Federal Environmental Ministry initially favoured the Federal Environmental Protection Agency (Umweltbundesamt) as the responsible body, industry sought a self-administering model to be run by industry. The compromise ultimately includes substantial demands made by industry. Pursuant to an ordinance (*Umweltauditgesetz-Beleihungsverordnung*) of 18 December 1995 enacted by the Federal Environmental Ministry, the Deutsche Akkreditierungs- und Zulassungsgesellschaft für Umweltgutachter mbH (DAU) adopts the task of determining the eligibility of environmental experts. The DAU is made up of groups representing the interests of industry and groups representing the interests of environmental experts. The groups representing the interests of industry cited as the central argument for this model the fact that ecoauditing would have to be administered by industry because of the voluntary participation of enterprises. However, ecoauditing was not conceived primarily as a marketing instrument for industry or even as a means of creating employment for environmental experts; rather, based on Article 130(s) of the EC treaty, ecoauditing should primarily serve environmental protection. The fact that these public interests should be considered dominant shows that the eligibility decision is not an internal matter for industry and that these tasks should have been granted to a public authority (Lübbe-Wolff 1996a: 219).

The self-administering model also conforms to the provisions of the Ecoauditing Law, according to which the chambers of industry and commerce and the chambers of crafts are entrusted with the task of registering the examined locations of enterprises. For the reasons mentioned above, the transfer to a public authority of this registration would also be preferable (Lübbe-Wolff 1996a: 224).

Even if the dominance of the self-administering organisation is justified from an environmental-protection perspective by the argument that this results in a greater willingness of enterprises to participate in audits, in the authors' opinion this could lead to a pyrrhic victory not only for the environment but also for the participating enterprises. The right of validated and registered enterprises to obtain a certificate declaring their participation, which offers the economic incentive to participate in ecoauditing, does not in itself help the enterprise. The right will only become important as a parameter of competition if the consuming public is convinced of the seriousness of the audit procedure and the impartiality of the experts involved. As a result of the present auditing system not only the environment but also enterprises which follow ecologically oriented environmental policies could suffer disadvantages. Whether this sceptical assessment of the auditing is justified will have to be proved by a careful evaluation of the auditing procedure, and in particular through an examination of the quality of published environmental certificates declaring participation.

Conclusions

The result of our attempt to assess the development of new environmental instruments in Germany is sobering. The 'old' state interventionist and command and control regulation instruments are no longer in fashion given changed social framework conditions. But effective alternatives, new means of regulating the relationship between man and nature, have not been found. Economic instruments (if one wants to refer to them as 'new'), although comprehensively discussed in view of their advantages, are not implemented in German law to such an extent that significant positive effects can be determined in practice. Even economists are no longer convinced of their suitability as substitutes for command and control law instruments. But they could provide a supplemental role. During the search for 'new instruments' the improvement and maintenance of 'old' ones must be given the same priority.

Environmental quality depends upon the politically desired and stipulated objectives even in the case of market-based regulations. As with emission levels in command and control regulation, economic incentives can be arranged so that they contribute only insignificantly to behavioural changes. Until there is a political willingness to stipulate objectives for overall emissions and overall waste amounts, and to base the level of environmental taxes on these objectives, environmental taxation will remain insufficient. In view of the prevailing fixation on improving national competitive advantages in the world market, such political objectives can not be expected (see Golub 1998). Because of the lack of international environmental regulation, a race to reduce environmental standards and international eco-dumping is more likely. A tendency towards an environmental law counter-reform is evidenced in Germany by the erosion of the procedural and participation rights of groups and the public, as well as by the reduction of legal-protection possibilities. If environmental standards are to be eroded it is not a good time for experiments with new instruments.

Since the mid-1980s the decisive innovations in environmental law in Germany have come from the EC. However, from an environmental perspective the implementation of EC law in Germany was often done in a manner that was not very efficient and indeed somewhat dubious. Important innovative instruments of the EC, such as the Directive on the Freedom of Access to Information on the Environment, are based on a functioning society which incorporates 'environmental interests' in administrative procedures and permits the general public to influence decisions. In view of the dominance of the discourse regarding economic problems and a more competitive situation (however real these factors may be), environmental protection is given low priority in comparison to employment arguments and economic interests in decision-making which requires the balancing of these interests. This means that control by society loses importance if the scope for environmental protection diminishes. It is necessary at least to substitute the 'weak' controls by 'general public opinion' with institutionalised rights to be heard and to object. If the scope for national environmental protection is dwindling because of international competition it is obviously necessary to secure minimum standards internationally. These could be introduced at supranational level directly as environmental protection standards. In addition, however, adoption of social minimum standards and economic-political regulation by the EC would increase the scope for environmental protection in the nation-states. The responsibility for environmental protection should therefore not be delegated; rather, the discussion of environmental law instruments must involve more closely the necessary economic and social framework conditions.

Notes

1 The WHG also contained planning elements to be enacted by the German states (*Bundesländer*), namely guidelines to create the framework for water-supply measures.
2 The advisors on the drafting of the Chemical Law and the 6th Amendment to EC Directive 67/548 influenced each other simultaneously, which ultimately led to the enactment of the Amendment to the Directive on 18 September 1979 and of the Chemical Law.
3 'Danger' requires at least a threat of a violation of the law.
4 During discussion of the WHG the aspect of providing a totality of services for the public had already been raised as a motive for effective environmental regulation.
5 For the different forms of planning, see Wegener 1975: 365; Albers 1969: 10; Wagener 1970: 93; Schefer 1979: 98; Breuer 1970: 101. For water law planning, see Breuer 1987: 253. It should be stressed here that global management at state level by the federal government was partially delegated to municipal levels (for example standardisation in construction planning law). As with central state planning, local planning was supposed to be extended, improved and simultaneously integrated into state (*Landespläne*) and federal (*Raumordnungspläne*) land planning.
6 Within society there were efforts to expand the rights of participation into, for example, the participation rights of employees or within the internal management of schools or tertiary-education establishments.
7 A paradigm which today still dominates parts of the environmental discussion, in particular the discussion about the economic advantages of new environmental technology.
8 These elements show only diffuse basic assumptions of a theory within the ecological movement. In fact the details of this theory vary significantly and the theory is in part very sophisticated.
9 These basic threads found greatest expression in Habermas's concept of the living world as a stronghold of communicative reason which should be protected against colonisation by the state and economic system (Habermas 1981).
10 Since then, however, a counter-reaction has been witnessed in the discussion of environmental groups. As the command and control regulation instruments have been cut back without economic instruments having been introduced, the classical instruments are being defended.
11 It appears to be undisputed that an internationalisation of the markets has taken place. It is, however, disputed whether this can be characterised as globalisation, that is, as unrestricted competition across the whole globe. The opponents of

this view assume the development of competition between three blocks which have formed around Japan, the US and Europe, but whose economic relationships mostly exist within the blocks, for example within the EU (Heise 1996: 17; Göll 1995: 545).

12 Moreover, in Germany efforts are being made towards establishing a comprehensive environmental law code (see Bender *et al.* 1995: 6 for many examples). In particular, enforcement could profit from a comprehensive codification of environmental law (Koch 1996: 217).

13 The Directive on the Freedom of Access to Information on the Environment cannot easily be slotted into any of the categories but supports many of the instruments.

14 In respect of subsidies and public contracts as instruments of environmental management (which are not dealt with here), see Bender *et al.* 1995: 53.

15 Nevertheless, the state retains an important role even in the realm of new instruments. A clear indication of this was the adoption of the state's objective of environmental protection (*Umweltschutz als Staatszielbestimmung*) in the German Constitution, Article 20a (Murswiek 1996b: 222; Kloepfer 1996: 72).

16 In respect of these positive external effects in the light of enterprises' research efforts, see Cremer 1995: 23.

17 To this extent it is often difficult to quantify the damages financially. It is even more difficult to attribute these damages to the entity which caused them (Endres 1994: 13; Bender *et al.* 1995: 51; Keppler 1994: 121).

18 In the economic theory literature a negotiating model based on ownership of rights is comprehensively discussed, linked to the so-called Coase-Theorem (Coase 1960). This involves polluters and those who suffer external effects negotiating over the level of the effects. This is preceded by a command and control political policy decision of principle, which allocates to one of the two parties the ownership right to the 'environment' resource. This model is, however, unsuitable for environmental-political practice for various reasons (Endres 1994: 33, 41).

The discussion of environmental taxes is not held only at a political level in Germany. Whether constitutional law permits the imposition of various environmental taxes is also vehemently disputed. As some environmental taxes have already been introduced this question has also been considered by various German courts, including the Federal Constitutional Court. However, this is not the place to describe the current dispute and the relevant legal precedents. It should simply be noted that, in the authors' opinion, in a 1995 decision (about a tax on the removal of water implemented in the states of Baden-Württemberg and Hesse) the Federal Constitutional Court paved the way for an 'ecological resource economy' (see Murswiek 1996a: 417).

19 However, there are practical obstacles to the legal implementation of the 'Pigou tax' and environmental-political concerns because of its orientation to Pareto-optimal rather than actual damages (Endres 1994: 95–6; Keppler 1994: 128).

20 In 1995 the Federal Administrative Court held that a municipal one-way waste tax (city of Kassel) constitutes a consumption tax, the levying of which by a municipality is constitutionally permissible as long as there is no similar federal tax (i.e. no such tax exists in the particular case).

21 Approval by preferably all EU states is (naturally) favoured (DGB 1996a: 5).

22 The revenue is to be used to help finance tasks which are difficult to insure in the form of a high federal subsidy to the Federal Employment Office. The tax is to be introduced in 1997, increased in stages until the year 2000 and then evaluated. The levels of tax will be set afresh each year, with the increase based on

revenue objectives. The DGB expects revenues of approximately DM14 million in the first year and around DM30 million in 2000 (see DGB 1996a: 6; 1996b: 2).

23 The most recent being the minister-president of Lower Saxony and SPD spokesman on politics and the economy, Gerhard Schröder.

24 Service industries will hardly be affected.

25 Currently DM49 per person per year.

26 A large proportion of plastic waste is disposed of overseas (*Frankfurter Rundschau*, 25 June 1996: 13). For a generally positive assessment of the packaging regulation, see Koch 1996: 218.

27 In particular, concerns relating to competition law have been raised about the functioning of the market-economy components of the model.

28 For a different view, see Wasmeier 1992: 220. See also Endres 1994: 104, who sees no enforcement advantages.

29 Bender (1986: 335) calls for either state liability or a fund solution. Kinkel (1989: 293) favours the fund solution and thinks that the state's financial contribution should be limited to providing the starting capital.

30 For example, the law relating to the construction of the Wismar-Ost to Wismar-West sector of the A20 road of 19 April 1994.

31 There is, however, a further discretionary provision to carry out an environmental impact assessment (EIA) with public participation within the context of a *Raumordnungsverfahren*. In light of the regulatory powers of the federal states the relationship between the various provisions is unclear.

32 That the EIA is linked to procedure and not the material effect of the plan is the subject of controversy between the EU Commission and the federal German government (Paetow 1996: 57).

33 The draft of the law provided by the Federal Environmental Ministry dated 29 November 1995 followed the drafts of the federal government on the amendment to the BImSchG, the 4th *Bundesimmissionsschutzverordnung*, the *Verwaltungsverfahrensgesetz* and the *Verwaltungsgerichtsordnung*, BR-Drs. 27/96 of 12 January 1996 and 29/96, 30/96 and 31/96 of 19 January 1996. The cited drafts may have been implemented as laws at the time of publication but, due to resistance from the upper house of the German parliament (the Bundesrat), only in a modified form.

34 See the objectives of the aforementioned government drafts, BR-Drs. 27/96, 29/96, 30/96 and 31/96.

35 The European Court emphasises this viewpoint particularly frequently in competition law (Everling 1993: 215).

36 Besides enforcement before national courts with the help of a generous ability to litigate granted by Community law, it is also possible to provide the Commission with information to follow the Article 169 of the EC treaty route.

37 Council Directive of 7 June 1990 regarding free access to information about the environment (90/313/EC), EC OJ 1990 No. L158, p. 56. The deadline for implementation was 31 December 1992. The German Law came into force on 16 July 1994.

38 See Scherzberg 1994: 733 for a critical viewpoint. See also the less restrictive drafts from the Federal Environmental Ministry: *Sachverständigenrat für Umweltfragen Umweltgutachten 1994*, BT-Drs. 12/6995, Tz. 560.

39 For the history of access to official documents, see Engel 1993: 11, and for recent political initiatives to expand the rights of access to documents, see Theurer 1996: 326, n.2.

40 The 'king's route to environmental protection' provides improved public information (Führ 1994: 453, 468).
41 One author (Stockburger 1991: 317) has described the public as performing the work of an 'environmental-stasi'.
42 In addition, the Ecoaudit Regulation gives rise to interpretation problems (see Lübbe-Wolff 1996a: 219). In Germany the question whether the validation of the environmental declaration by the admitted environmental expert constitutes actual compliance with the applicable environmental provisions at the location is very much disputed. For views in favour, see Schottelius 1995: 1551; Falke 1995: 5; Lübbe-Wolff 1994: 369. For views against, see Sellner and Schnutenhaus 1993: 930; Förschle *et al.* 1994: 1099; Schnutenhaus 1995: 9; Schneider 1995: 378. For criticism of the lack of protection for commercial confidential interests in the Regulation, see Martens and Moufang 1996: 246.
43 Even between the Bundesrat (upper house of parliament) and the Bundestag (lower house of parliament) there were individual points of dispute, as a result of which the Bundesrat called upon the mediation committee of the German Bundestag and Bundesrat.

References

X (1996) *Aktualisierte Erklärung der deutschen Wirtschaft zur Klimavorsorge vom 27 March*: 3 (Bonn: Bundesregierung).

Albers, G. (1969) 'Über das Wesen der räumlichen Planung', *Stadtbauwelt 1969* 2(1): 10–13.

Altvater, E. (1992) *Die Zukunft des Marktes* (Münster: Dampfboot).

Behrens, P. (1986) *Die ökonomischen Grundlagen des Rechts* (Tübingen: Mohr).

Bender, B. (1986) 'Zur staatshaftungsrechtlichen Problematik der Waldschäden' *Verwaltungsarchiv* 77(4): 335–71.

Bender, B., Sparwasser, R. and Engel, R. (1995) *Umweltrecht*, 3rd edn, (Heidelberg: Müller).

Beschluß der Bundesregierung (1994) *Zur Verminderung der CO_2-Emmissionen und anderer Treibhausgasemmissionen in der Bundesrepublik Deutschland auf der Grundlage des dritten Berichts der Interministeriellen Arbeitsgruppe 'CO_2-Reduktion'*, Bundestags-Drucksache (BT-Drs.) 12/8557, 5 October.

Bidner, C. (1994) 'Über den Zugang zu Umweltinformationen in Österreich', *Umwelt und Planungsrecht* 14(11–12): 408–15.

Bohne, E. (1981) *Der informale Rechtsstaat* (Berlin: Duncker & Humblot).

Brandt, W. (1961) 'Regierungsprogramm in Broschüre zum Regierungsprogramm 1961' (Bonn: SPD).

Breuer, R. (1970) 'Selbstbindung des Gesetzgebers durch Programm- und Plangesetze?' *Deutsches Verwaltungsblatt* 85(3): 101–5.

—— (1987) *Öffentliches und privates Wasserrecht*, 2nd edn (Munich: Beck).

—— (1990) 'EG-Richtlinien und deutsches Wasserrecht', *Wirtschaft und Verwaltung* 2: 79–117.

—— (1992) 'Umweltrechtliche und wirtschaftslenkende Abgaben im europäischen Binnenmarkt', *Deutsches Verwaltungsblatt* 107(8): 485–96.

—— (1995) 'Umweltschutzrecht, besonderes Verwaltungsrecht', in E. Schmidt-Aßmann (ed.) *Besonderes Verwaltungsrecht* (Berlin/New York: de Gruyter).

Bundesverband der Deutschen Industrie (1996) *Report 24/96*, 27 March.

Cansier, D. (1994) 'Gefahrenabwehr und Risikovorsorge im Umweltschutz und der Spielraum für ökonomische Instrumente', *Neue Zeitschrift für Verwaltungsrecht* 13(7): 642–7.

Coase, R. (1960) 'The problem of social costs', *Journal of Law and Economics* 3(3): 215–38.

Cremer, W. (1995) *Forschungssubventionen im Lichte des EGV* (Baden-Baden: Nomos).

Demmke, C. (1994) *Die Implementation von EG-Umweltpolitik in den Mitgliedstaaten* (Baden-Baden: Nomos).

DGB (1996a) *Eckpunkte*, 4 June.

—— (1996b) *Positionspapier*, 4 June.

Dose, N. (1990) 'Durch Abgaben zu mehr Umweltschutz?' *Zeitschrift für Umweltpolitik und Umweltrecht* 13(4): 365–78.

Draeger, K. and Wolf, F. (1995) 'Teure Energie, billige Arbeit?', *Blätter für deutsche und internationale Politik* 15(6): 1071–85.

Ehrig, D. (1989) *Keynes und die Globalsteuerung und die Stabilitätspolitik in der Bundesrepublik Deutschland* (Frankfurt: Lang).

Endres, A. (1994) *Umweltökonomie* (Darmstadt: Wissenschaftliche Buchgesellschaft).

Endres, A., Rehbinder, E. and Schwarze, R. (eds) (1993) *Umweltzertifikate und Kompensationslösungen aus ökonomischer und Juristischer Sicht* (Bonn: Economia).

Engel, C. (1993) *Akteneinsicht und Recht auf Informationen über Umweltbezogene Daten* (Baden Baden: Nomos).

Erichsen, H. (1992) 'Bedeutungsverlust der Verwaltung im Gefüge staatlicher Gewaltenteilung', *Neue Zeitschrift für Verwaltungsrecht* 11(6): 409–12.

Everling, U. (1992) 'Umsetzung von Umweltrichtlinien durch normkonkretisierende Verwaltungsanweisungen', *Recht der internationalen Wirtschaft* 38(5): 379–85.

—— (1993) 'Durchführung und Umsetzung des europäischen Gemeinschaftsrechts im Bereich des Umweltschutzes unter Berücksichtigung der Rechtsprechung des EuGH', *Neue Zeitschrift für Verwaltungsrecht* 12(3): 209–16.

Ewringmann, D. (1994) 'Ökologische Steuerreform?' *Zeitschrift für Umweltpolitik und Umweltrecht* 17(1): 43–56.

Falke, J. (1995) 'Umwelt-Audit-Verordnung', *Zeitschrift für Umweltrecht* 6(1): 4–9.

Fisahn, A. (1996) 'Kampf gegen Windmühlen?' *Neue Justiz* 50(2): 63–70.

Fluck, J. and Theurer, A. (1996) *Umweltinformationsrecht* (Heidelberg: Müller).

Förschle, G., Herrmann, S. and Mandler, U. (1994) 'Umwelt-Audits', *Der Betrieb* 47(22): 1093–100.

Friedman, M. (1962) *Capitalism and Freedom* (Stuttgart: Seewald).

Führ, M. (1994) 'Praktisches unternehmerisches Handeln', *Zeitschrift für Umweltpolitik und Umweltrecht* 17(4): 445–72.

Gawel, E. and Ewringmann, D. (1994) 'Die Kompensationsregel der TA Luft', *Natur und Recht* 16(3): 120–5.

Gellermann, M. (1994) *Beeinflussung des Bundesdeutschen Rechts durch Richtlinien der EG – dargestellt am Beispiel des europäischen Umweltrechts* (Cologne: Heymann).

Gieseke, P. (1992) *Wasserhaushaltsgesetz*, 6th edn (Munich: Beck).

Göll, E. (1995) 'Nafta oder die Regionalisierung des Nord-Süd Konflikts', *Neue Gesellschaft – Frankfurter Hefte 1995* 42(6): 545–67.

Göll, E. and Schuster, J. (1995) 'Rückkehr zum Primat der Politik', in E. Bulmahn, von Oertzen, P. and Schuster, J. (eds) *Jenseits der Öko-Steuern* (Dortmund: SPW).

Golub, J. (ed.) (1998) *Global Competition and EU Environmental Policy* (London: Routledge).

Goßler, C. (1990) 'Mehr Markt im Luftreinhalterecht', *Umwelt- und Planungsrecht* 10(7): 255–8.

Habermas, J. (1981) *Kommunikatives Handeln*, vol. II, (Frankfurt: Suhrkamp).

Hansmeyer, K. and Schneider, H. (1992) *Umweltpolitik*, 2nd edn (Göttingen: Vandenhoek & Rupprecht).

Heise, A. (1996) 'Der Mythos vom "Sachzwang Weltmarkt"', *Internationale Politik und Gesellschaft* 22(1): 17–35.

Hendler, R. (1990) 'Umweltabgaben und Steuerstaatsdoktrin', *Archiv des öffentlichen Rechts* 115(4): 577–609.

Hennecke, H. (1991) 'Informelles Verwaltungshandeln im Wirtschaftsverwaltungs- und Umweltrecht', *Natur und Recht* 13(6): 267–71.

Herr, H. (1988) 'Wege zur Theorie einer monetären Produktionswirtschaft – der keynesianische Fundamentalismus', *Ökonomie und Gesellschaft* 6: 66–98.

Heyvaert, V. (1997) 'Access to information in a deregulated environment', in U. Collier (ed.) *Deregulation in the European Union: Environmental Perspectives* (London: Routledge).

Hoffmann, K. (1988) 'Kampf mit allen Mitteln', *Manager Magazin* 3: 148–50.

Horbach, J. (1992) *Neue politische Ökonomie und Umweltpolitik* (Frankfurt/New York: Campus).

Hübner, K. and Stanger, M. (1986) 'Kleine und große Krisen', in S. P. W. Prokla (ed.) *Kontroversen zur Krisentheorie* (Hamburg: VSA-Verlag).

Jass, M. (1990) *Erfolgskontrolle des Abwasserabgabengesetzes* (Frankfurt/New York: Campus).

Kalmbach, P. and Kurz, H. (1983) 'Klassik, Neoklassik und Neuklassik', *Ökonomie und Gesellschaft* 1: 57–92.

Keppler, J. (1994) 'Internalisierung der externen Kosten der Energie', in M. Vohrer (ed.) *Ökologische Marktwirtschaft in Europa* (Baden-Baden: Nomos).

Kinkel, K. (1989) 'Möglichkeiten und Grenzen der Bewältigung von umwelttypischen Distanz- und Summationsschäden', *Zeitschrift für Rechtspolitik* 22(8): 293–8.

Kirsch, G. (1993) *Neue politische Ökonomie* (Berlin: Werner).

Klocke, U. (1995) *Klimaschutz durch ökonomische Instrumente* (Baden-Baden: Nomos).

Klodt, H. and Stehn, J. (1994) *Standort Deutschland* (Tübingen: Mohr).

Kloepfer, M. (1994) *Zur Geschichte des deutschen Umweltrechts* (Berlin: Duncker & Humblodt).

—— (1996) 'Umweltschutz als Verfassungsrecht: zum neuen Art. 20a GG', *Deutsches Verwaltungsblatt* 111(2): 73–80.

Koch, H. (1996) 'Vereinfachung des materiellen Umweltrechts', *Neue Zeitschrift für Verwaltungsrecht* 15(3): 215.

Köck, W. (1993) 'Umweltabgaben – quo vadis', *Juristenzeitung* 48(2): 59–67.

Körner, H. (1986) *Die Zukunft der Globalsteuerung, Karl Schiller zum 75. Geburtstag* (Bern/Stuttgart: Haupt).

Lahl, U. (1993) 'Das programmierte Vollzugsdefizit', *Zeitschrift für Umweltrecht* 4(4): 249–54.

Länderrat von Bündnis 90/Die Grünen (1995) 'Eckpunkte für den Einstieg in eine ökologisch-soziale Steuerreform', September.

Lübbe-Wolff, G. (1993) 'Vollzugsprobleme der Umweltverwaltung', *Natur und Recht* 15(5): 217–29.
—— (1994) 'Die EG-Verordnung zum Umwelt-Audit', *Deutsches Verwaltungsblatt* 109(7): 361–74.
—— (1995) 'Beschleunigung von Genehmigungsverfahren auf Kosten des Umweltschutzes', *Zeitschrift für Umweltrecht* 6(2): 57–62.
—— (1996a) 'Das Umweltauditgesetz', *Natur und Recht* 18(5): 217–27.
—— (1996b) *Modernisierung des Umweltordnungsrecht* (Bonn: Economica).
Luhmann, N. (1984) *Soziale Systeme* (Frankfurt: Suhrkamp).
—— (1986) *Ökologische Kommunikation* (Opladen: Westdeutscher).
Lütkes, S. (1996) 'Das Umweltauditgesetz – UAG', *Neue Zeitschrift für Verwaltungsrecht* 15(3): 230–5.
Martens, C., and Moufang, O. (1996) 'Kritische Aspekte bei der praktischen Durchführung der Öko-Audit-Verordnung', *Neue Zeitschrift für Verwaltungsrecht* 15(3): 246–7.
Mayntz, R. (1978) *Vollzugsprobleme der Umweltpolitik* (Stuttgart: Gustav Fischer).
Mies, M. (1988) *Patriachat und Kapital* (Zurich: Rotpunkt).
—— (1996a) 'Ein Schritt in Richtung auf ein ökologisches Recht', *Neue Zeitschrift für Verwaltungsrecht* 15(5): 417–21.
Murswiek, D. (1996b) 'Staatsziel Umweltschutz (Art. 20a GG)', *Neue Zeitschrift für Verwaltungsrecht* 15(5): 222–30.
Paetow, S. (1996b) 'Beschleunigungsmaßnahmen bei der Fernstraßenplanung', *Zeitschrift für Umweltrecht* 7(2): 57–60.
Pigou, A. (1952) *The Economics of Welfare*, 4th edn (London: Macmillan); 1st edn 1920.
Rehbinder, E. (1992) 'Haftung und Versicherung für Umweltschäder aus ökonomischer Sicht', in A. Endres, E. Rehbinder and R. Schwarze (eds) *Umweltzertifikate und Kompensationslösungen aus ökonomischer und Juristischer Sicht* (Bonn: Economia).
Röger, R. (1995) *Umweltinformationsgesetz* (Cologne: Heymann).
Rohde, A. (1990) 'Möglichkeiten der Deregulierung in der Umweltpolitik', in P. Oberender and M. Streit (eds) *Soziale und ökologische Ordnungspolitik in der Marktwirtschaft* (Baden Baden: Nomos).
Schefer, A. (1979) 'Ein neuer Abschnitt der Raumordnung', *Deutsches Verwaltungsblatt* 85(2): 98–105.
Scherzberg, A. (1994) 'Freedom of information – Deutsch gewendet: das neue Umweltinformationsgesetz', *Deutsches Verwaltungsblatt* 110(14): 733–45.
Schlabach, E. (1990) 'Das 3. Änderungsgesetz zum Bundes-Immissionsschutzgesetz', *Umwelt- und Planungsrecht* 10(7): 250–5.
Schmidt-Aßmann, E. (1993) 'Deutsches und europäisches Verwaltungsrecht', *Deutsches Verwaltungsblatt* 108(17): 924–36.
Schneider, J. (1995) 'Öko-Audit als Scharnier in einer ganzheitlichen Regulierungsstrategie', *Die Verwaltung* 28(3): 361–88.
Schnutenhaus, J. (1995) 'Die Umsetzung der Öko-Audit-Verordnung in Deutschland', *Zeitschrift für Umweltrecht* 6(1): 9–13.
Schomerus, T., Schrader, C. and Wegener, B. (1995) *Umweltinformationsgesetz* (Baden-Baden: Nomos).
Schottelius, D. (1995) 'Das EG-Umwelt-Audit als Gesamtsystem', *Betriebsberater* 50(31): 1549–53.

Sellner, D. and Schnutenhaus, J. (1993) 'Umweltmanagement und Umweltbetriebsprüfung ("Umwelt-Audit") – ein Wirksames, nicht ordnungsrechtliches System des betrieblichen Umweltschutzes?' *Neue Zeitschrift für Verwaltungsrecht* 12(10): 928–34.

SPD (1995) Party Conference, 14–17 November, Mannheim.

Steinberg, R. (1993) *Fachplanung* (Baden-Baden: Nomos).

—— (1995) 'Zulassung von Industrieanlagen im deutschen und europäischen Recht', *Neue Zeitschrift für Verwaltungsrecht* 14(3): 209–19.

Stockburger, B. (1991) 'Instrumente zur Verwirklichung einer "gläsernen Unternehmerumwelt"', *Wirtschaftsverwaltung und Umweltrecht* 6: 315–27.

Theurer, A. (1996) 'Der Zugang zu Umweltinformationen aufgrund des Umweltinformationsgesetzes (UIG)', *Neue Zeitschrift für Verwaltungsrecht* 15(4): 326–33.

Turiaux, A. (1996) *Umweltinformationsgesetz* (Munich: Beck).

Umweltprogramm der Bundesregierung (1971) BT-Drs. VI/2710, 29 September.

VG (1995) 'Verwaltungsgericht Gelsenkirchen', *Natur und Recht* 17(3): 158.

Vogel, H. (1979) 'Zur Diskussion um die Normenflut', *Juristenzeitung* 34(6): 321–3.

Voigt, R. (1980) *Verrechtlichung* (Frankfurt: Suhrkamp).

Wagener, F. (1970) 'Von der Raumplanung zur Entwicklungsplanung', *Deutsches Verwaltungsblatt* 85(2): 93–7.

Wagner, H. (1995) 'Effizienz des Ordnungsrechts für den Umweltschutz', *Neue Zeitschrift für Verwaltungsrecht* 14(11): 1046–52.

Wasmeier, M. (1992) 'Marktfähige Emissionslizenzen – das Zertifikursmodell und seine Umsetzung in den USA', *Natur und Recht* 14(5): 219–26.

Wegener, G. (1975) 'Von der Entwicklungsplanung zur Aufgabenplanung', *Die öffentliche Verwaltung* 28(11): 365–70.

Wicke, L. (1991) *Umweltökonomie und Umweltpolitik* (Munich: Vahlen).

Winter, G. (1985) 'Bartering Rationality in Regulation', *Law and Society Review* 18(2): 210–15.

4

NEW ENVIRONMENTAL POLICY INSTRUMENTS IN THE NETHERLANDS

Duncan Liefferink[1]

Introduction

Dutch environmental policy has a certain reputation for being comparatively innovative. In particular, the National Environmental Policy Plan (NMP 1988–9) has been hailed by foreign observers as one of the first and most comprehensive policy programmes towards sustainable development (e.g. Weale 1992; Wallace 1995). Such judgements have to do not only with the ambitious objectives contained in the plan, but also with the new approach to environmental policy-making that it conveys. The ultimate goal of this approach, as the plan quite explicitly states, is the ecological transformation of society. In order to achieve this goal the plan relies heavily on a new generation of policy instruments, based particularly on communication, negotiation and consensus between all parties involved, public as well as private actors.

The emergence of this new generation of policy instruments is often associated with the notion of ecological modernisation. According to the theory of ecological modernisation, which can itself be placed in the broader framework of theories of reflexive modernisation,[2] the environmental challenge will lead to a new phase in the development of modern, industrial society, characterised by the ecological restructuring of state, market and technology (Huber 1982; Spaargaren and Mol 1992). In the political realm, as Jänicke (1993) argues, the failure of the traditional dirigiste state to cope with environmental problems will lead to a stronger reliance on other actors and a more 'stage-setting' role for state institutions. As regards the use of policy instruments, this may express itself in a shift from a 'command and control' approach to both economic and communicative instruments.

The Netherlands indeed appears to be moving in the forefront of this development (Weale 1992; Spaargaren and Mol 1992; Mol 1995; Hajer

1995; Le Blansch 1996; Liefferink and Mol 1997), but an interesting feature of the modernisation of Dutch environmental policy is the focus on communication and negotiation. As will be discussed in the next section, an effective system of environmental charges in the field of water pollution was established in the Netherlands as early as 1970. Generally speaking, however, charges remained of secondary importance in comparison with traditional command and control regulation and never really developed into a second generation of policy instruments, as they did in some other countries. Instead, a fundamental reorientation of environmental policy, starting in the 1980s, led to a shift from (predominantly) direct regulation directly to communicative or third-generation instruments.[3]

In the subsequent section the basic ideas behind the innovations of the last decade will be sketched. It will be shown that the National Environmental Policy Plan, published in 1989, should in fact be seen as an important step in an ongoing process rather than as its starting point. Then we will deal with the two most important fields of application of the communicative approach: industrial pollution and regional management. In both cases, the working of the policy system in practice will be briefly described, followed by a critical examination of pros and cons and some theoretical considerations.

In the concluding section the findings of the chapter will be summarised. Among other things, we will return briefly to the link between the evolution of Dutch environmental policy during the last decade and the notion of ecological modernisation.

Economic instruments and the exceptional case of surface-water pollution

Until the early 1980s direct regulation was the dominant form of environmental policy-making in the Netherlands. For instance, the basic laws on air pollution (1970), sea-water pollution (1975), chemical waste (1976), waste (1977) and noise (1979) are all based on various types of prohibition orders and product and process requirements, coupled with a system of licenses or registration. Most of these laws do not establish standards themselves, but empower the government to do so with the help of more specific decrees and regulations. Thus, emission standards were set at the national level and formed the basis for licenses, usually granted by the provincial or local authorities (or, in the case of water pollution, by the water boards – see below). Environmental-quality standards were not part of this system. They appeared only in non-binding policy programmes. The initial absence of a legal basis for quality standards explains why implementation of EC Directives on air-quality standards was long delayed in the Netherlands (Bennett 1991: 73–8).

The Surface-water Pollution Act (*Wet Verontreiniging Oppervlaktewateren*) was adopted in 1969 and came into force in 1970. The Act forbids the discharge of waste water unless a licence is obtained. In this sense, it is in line with the other basic environmental laws established in the Netherlands during the 1970s. The Surface-water Pollution Act is distinguished from the other first-generation pollution laws, however, by the levy which is imposed on all polluters, households as well as industry. This system of effluent charges is often depicted as a success story (most explicitly by Bressers 1983b, 1995; Bressers and Schuddeboom 1994). Considering the impact of the system on water quality in the Netherlands, this view is undoubtedly correct. Shortly after the inception of the levy the amount of organic substances discharged by industry started to decline sharply. Parallel to this, public sewage capacity rapidly increased. Apart from the environmental impact, the cost-effectiveness of the charge scheme also turned out to be comparatively high (Andersen 1994).

The unequivocal success of this early example of an environmental tax might suggest the considerable popularity of this type of instrument in the Netherlands. As will be shown, however, the Netherlands can in fact hardly be considered a pioneer in the application of market-based policy instruments. A number of specific features of the surface-water case, particularly of an institutional kind, may account for its exceptional character in the context of Dutch environmental policy.

The basis for calculation of the water charge is the amount of pollution discharged, particularly the amount of organic substances. The effluent charge was designed as a pure revenue charge, intended for financing the building and operation of public sewage treatment plants. A central role in the implementation of the policy is played by the water boards (*waterschappen*).[4] These are semi-autonomous, semi-democratic institutions dating back to the Middle Ages. Their traditional activities are related to water quantity (including the management of reclaimed land), but in recent decades they have been assigned tasks in the field of water quality as well. In most cases, the water boards are responsible for collecting effluent charges, with the help of which they operate sewage plants in their region. As the costs of sewage treatment may differ among regions – due, for instance, to variations in population density and industrial structure – the levy is not equally high throughout the country.

Apart from the construction of a large number of sewage plants financed by it during the 1970s, the most important impact of the Surface-water Pollution Act has been the sharp decrease in industrial discharges. As Bressers demonstrated in 1983, the effluent charge in fact had a strong regulative effect. Due to its considerable size, immediately related to the amount of pollution, and because it was imposed on emissions directly into surface waters as well as on discharges to the sewerage system, the levy worked as a forceful incentive for firms to take their own measures to reduce emissions. Such measures could involve either the construction of in-house sewage treatment or the shift to cleaner technologies. The overall result was a decrease in total industrial organic discharges of about 50 per cent between 1970 and 1975, and 80 per cent between 1970 and 1986. According to Bressers' analysis, the decrease can largely be attributed to the effluent charge Bressers 1983a, 1983b). An interesting secondary effect was found some years later by Schuurman (1988). He showed that the massive efforts in water treatment by industry itself led to a considerable overcapacity of public sewage plants. This, in turn, contributed to the increase in the amount of the levy for the remaining dischargers and thus reinforced the incentive for firms to take their own measures.

Although Bressers (1995: 40) may be right in claiming that the conditions for the success of the effluent charge are not exclusively Dutch, two quite specific circumstances should be mentioned here. In the first place, as Bressers also points out (*ibid.*), it is essential that the authorities collecting the levy are themselves strongly dependent on the revenues for performing their tasks. This in itself is not unique and can, in principle, be the case for many revenue charges. The particular, historically determined institutional character of the water boards, however, made the Dutch effluent charges, to some extent, a special case. As mentioned above, the water boards have for centuries played a crucial role in the management of water quantity in the Netherlands – an issue of vital importance in the Low Countries. Each water board is governed by an elected body consisting of representatives of those directly interested in the board's activities and contributing financially to it. In 1990 the system of finance and representation was changed slightly, but at least until that time traditional water-quantity interests, particularly farmers, dominated the water boards. This circumstance, together with the depoliticised, technocratic orientation of the boards and their staff, made it relatively easy to impose significant costs on industrial polluters, who hardly had a say in the water boards anyway (Andersen 1994).

A second specific factor can be found in the somewhat confused distribution of competences in Dutch water management at large. Whereas the water boards take care of most of the smaller surface waters, the Directorate for Public Works (Rijkswaterstaat; the executive branch of the Ministry of Transport and Public Works) is responsible for the largest water courses, such as the North Sea and the great rivers. This directorate also collects levies for discharges into the waters it administers, but it does not operate any sewage plants of its own. Instead, the directorate's research institute (Rijksinstituut voor Zuivering van Afvalwater – RIZA) used the money to establish a programme for the development of control technologies for industry. The spreading of this know-how was facilitated by close cooperation between RIZA and the industrial sectors involved, and by a subsidy scheme, financed by the revenues of the levy (Andersen 1994).

As was pointed out at the beginning of this section, Dutch environmental policy in the 1970s and early 1980s relied heavily on command and control strategies. The approach chosen in the field of surface-water pollution can be regarded as the exception to this rule. Moreover, Andersen's analysis, in particular, gives rise to the observation that the specific institutional constellation around the issue of water pollution turned the effluent charges into a very special case indeed. At the same time, it should be noted that even this special case derived its success at least partly from its combination with other policy instruments. On the one hand, the charges were still linked to the traditional permit system used throughout Dutch environmental policy at the time. On the other hand, the advisory and supportive role of RIZA points to the importance of more cooperative forms of interaction between the state and polluters in the Netherlands. Bressers (1983b: ch. 8; 1995: 39) recognises the significance of negotiation for the achievement of policy goals, particularly in the case of the discharge of heavy metals into surface water, where charges were relatively lower than for organic pollution.

To emphasise the exceptional character of the effluent charge is not to say that economic instruments are applied in no other field in the Netherlands. In 1994 the total revenue of environmental charges in the Netherlands amounted to G6.9 billion, only about G1.9 billion of which was accounted for by the charges under the Surface-water Pollution Act. Another G3.0 billion consisted of pure revenue charges for the collection of household waste and the maintenance of the public sewerage system. The lion's share of these charges is unrelated to the pollution produced or the actual use of the services involved, and therefore does not have any regulative effect (Vermeulen 1994).[5] In the present context, the remaining part of the G6.9 billion mentioned is the most interesting.

Since 1972 there has been a small levy on fuels, the revenues of which were used for a fund to compensate for both damage and excessive abatement costs related to air pollution. From 1992 this levy was transformed into a user tax on fuels, generating some G2 billion annually. Two years later this tax was extended to water use, with an option for a further broadening. The revenues of these user taxes are not earmarked and flow into the general state budget. If, in the longer term, taxpayers will be compensated by the reduction of other taxes, the user taxes can be seen as a step towards shifting the tax burden from labour and capital to environment and resource use, sometimes referred to as the 'greening' or 'ecologising' of the tax system (Vermeulen 1994; WRR 1992: chs 1 and 11).

The recent carbon-dioxide (CO_2) tax follows a similar line of thought. Its revenues are not used for specific environmental purposes but returned to taxpayers in the form of lower income taxes and social contributions. Existing funds to subsidise, for example, research into higher energy efficiency or home insulation were not coupled with the CO_2 tax and were in fact even reduced in the period when the tax was introduced. Thus the CO_2

tax is revenue-neutral and should be seen as a regulative charge. Nevertheless, the establishment of the CO_2 tax in the Netherlands was a long and difficult process. In the early 1990s a number of scenarios for a tax to curb CO_2 emissions were seriously discussed (Steering Committee on Regulating Energy Taxes 1992; see also Jaarsma and Mol 1994). Yet a decision on the introduction of a national tax was postponed with a view to the ongoing negotiations about an EU-wide tax. The Dutch economy is strongly export-oriented and characterised by a relatively high share of often large, energy-intensive firms, particularly in the chemical, petrochemical and metals sectors. This explains the strong and successful opposition of industry to unilateral measures which might distort competitive conditions. Only when it appeared that no agreement in the EU context could be expected in the near future was the decision taken to introduce a national tax scheme granting exemptions to the energy-intensive industries. The tax that eventually took effect on 1 January 1996 applies only to small users, i.e. households and small and medium-size firms using less than 170,000 m^3 natural gas or 50,000 kW electricity per year. Large energy users were to be covered by negotiated agreements to improve energy efficiency. A considerable number of such agreements had already been concluded between various sectors and the state in the years before.

In addition to environmental taxes, a number of other financial instruments have been applied in the Netherlands (see also Vermeulen 1994; Bressers and Schuddeboom 1994). In the late 1980s an effective tax-differentiation scheme stimulated sales of 'clean' cars and unleaded petrol (Bennett 1991; Liefferink 1996). As in other countries, a gradual revival of traditional deposit-and-return systems can be observed, for instance for polyethylene (PET) bottles for soft drinks. In distinction to, for instance, Denmark, however, one-way beverage containers were not banned, nor was the use of standardised types of bottles imposed upon producers. As a consequence, no conflicts with foreign producers or Brussels occurred (these potential conflicts are discussed in Chapter 11). When one buys a new car in the Netherlands, moreover, an earmarked contribution has to be paid for the costs of final disposal and recycling. The reverse of this system existed until recently for refrigerators, where consumers received a bonus when they disposed of their old fridge to a controlled processor.

Considering only the *number* of environmental taxes, the Netherlands is neither a leader nor a laggard in comparison with other industrialised countries (Opschoor 1994). As we have seen, however, a considerable number of these are pure, fixed-tariff revenue charges, used to finance public services (waste collection, sewerage) and not intended to change the behaviour of polluters. As such, they can be compared with the levy to be paid for a new passport and can hardly be counted as belonging to a new generation of policy instruments. As for environmental charges actually designed to have an impact on polluting activities, these mainly related to the behaviour of the large and diffuse group of consumers. With the notable exception of the effluent charge, industry – and particularly the larger industrial sectors – found a relatively willing ear for its objections against regulative charges. In 1985, for instance, the Dutch employers' associations argued that environmental charges were acceptable only in as far as they were returned directly to the firms involved in the form of specific services (VNO and NCW 1985). Among the reasons for the reluctance on the part of the government to impose direct financial burdens on industry can be listed the small size of the country and the very open, export-oriented character of the Dutch economy, as well as the emphasis put on this aspect in Dutch politics. Rather than economic instruments, therefore, negotiation and cooperation between the state and industry have become the dominant alternative to traditional command and control strategies in the Netherlands. Obviously, consumers are more difficult to approach through negotiation. Moreover, they are hardly affected by international competition.[6]

The political context of environmental covenants: internalisation and target group policies

As was pointed out above, environmental policy in the 1970s and early 1980s was dominated by direct regulation. It should be noted, however, that in this period consultation and negotiation were also part and parcel of the process of policy-making and policy implementation. The Netherlands is a country with a strongly consensual policy style (Van Putten 1982). During the preparation of new laws and decrees, extensive consultations usually take place between the ministries involved and the actors to which the policy is addressed. In this process, branch organisations play a central role. The environmental field was generally no exception to this, even though the new Ministry of the Environment, founded in 1971, had to assert itself against various more established interests (see Van Tatenhove 1993). In addition, negotiation and accommodation to specific circumstances on a case-by-case basis was (and still is) an important component of licensing procedures. Most environmental licenses are issued by the local or provincial authorities.

Another characteristic of the first phase of the institutionalisation of environmental policy in the Netherlands was the strong sectorisation. After a decade of vigorous attempts to cope with a rapidly expanding range of ecological threats, it was hardly an exaggeration to say that every subfield of environmental policy had its own regulations. The coherence between the various subfields was poor and so was coordination with adjacent policy areas such as economic, agricultural or transport policy. Industry, moreover, complained about having to deal with various permits in parallel, each following its own procedures and time paths. In the early 1980s this burden became even more pressing because of stagnating economic growth. In short, both the effectiveness and efficiency of environmental policy were called into question.[7] It was the first Christian-Democrat/Liberal cabinet headed by Prime Minister Lubbers (1982–6) and particularly its minister for the environment, Pieter Winsemius, which took up the challenge to accomplish a far-reaching reorientation and integration of environmental policy. On the one hand, the concept of integration had an 'internal' component, referring to the connections between different environmental problems and policies. On the other hand, it entailed the 'external' integration with other areas and, as such, the institutional affirmation of environmental protection as a full and mature policy field (Van Tatenhove and Liefferink 1992; Van Tatenhove 1993; Leroy 1994).

Integration was to be achieved along two complementary lines, or 'tracks' (IMP-M 1985–9, 1986–90). The *effect-oriented* line started from the assessment of environmental quality and ecological risks. On the basis of that, targets and policies were formulated to protect environmental values. As such, effect-oriented policies set the framework for the *source-oriented* line, which focused on the prevention or limitation of pollution at the source. In the 1970s the latter aspect had been strongly emphasised. The two-track policy implied a certain reappraisal of the notion of environmental quality as the basis of environmental policy (Winsemius 1986: 38; Van Tatenhove 1993).

In order to overcome the drawbacks of the old fragmented approach, where problems could easily be shifted from one medium to the other, effect-oriented policies were to be organised around five environmental *themes*: acidification, eutrophication, diffusion of substances, disposal of waste and disturbance (including noise, odour, local air pollution, etc.). Later, climate change, dehydration and squandering were added. Apart from that, encompassing policies were formulated for natural areas with special ecological values. At the end of the 1980s the policies for special areas were extended to programmes encompassing a broad range of environmental, social and economic requirements in larger *regions*. This so-called region-oriented approach, which came to be seen as a third, spatial 'track' for integration, will be discussed in more detail below.

The problem of the fragmentation of policies with regard to sources of pollution was dealt with especially by formulating more coherent policies for the *target groups* of environmental policy, including various industrial sectors, traffic and transport, agriculture and households (consumers). One of the core objectives of this approach was the internalisation (*verinnerlijking*) of environmental responsibility with the target groups. Producers and consumers were to develop environmentally sound behaviour, not just as a result of government regulation, but also as an increasingly 'normal' element of daily practice. On the one hand, internalisation can be interpreted as a moral appeal to society, an aspect repeatedly stressed by the father of the concept, Winsemius. This is particularly well illustrated by the title of his political testament, 'Guest in one's own house' (Winsemius 1986). On the other hand, internalisation is a strategic goal, aiming at creating a broader basis for environmental measures and, eventually, at a fundamental shift in production and consumption cycles. Strengthening the role of actors in society in this process, in Winsemius's view, was to be achieved by more directly involving them in the making and implementation of environmental policies. Environmental charges did not really fit into this strategy. Although Winsemius was not against the use of economic instruments as such, he strongly emphasised the limitations to their use in practice (Winsemius 1986: 97–8). As a former director of a business consultancy firm, moreover, Winsemius was well aware of the industry's aversion to measures which might affect the (international) competitiveness of firms, including regulative charges (see above).[8] Particularly in the case of industry, therefore, the target-group approach predominantly took the form of negotiation and cooperation between the state and polluting sectors. In the decade since its inception the target-group approach has been widely applied in the Netherlands and resulted in about a hundred agreements, or covenants,[9] between the state and industry. In the next section we will examine how the strategy works in practice.

In a sense, the target-group approach is based on a type of exchange relationship that is reminiscent of neo-corporatist arrangements as they have long existed, for instance in social-economic policy and in some countries in the agricultural field (Cawson 1986; Frouws 1993). In exchange for substantial influence on the content of the policy, private actors commit themselves to the outcomes of the process and take over certain operational tasks, for instance those related to the establishment and control of internal environmental management systems. Negotiations in the environmental field, however, lack the tripartite structure of neo-corporatism in the social-economic field.[10] In addition, relations in the target-group approach are not as fixed as in corporatism. Instead, participation in the network is itself subject to negotiation (Le Blansch 1996).[11]

Furthermore, at first sight there seems to be a link between the target-group approach and deregulation as pursued particularly by the Lubbers Cabinet in the first years of its existence, i.e. in the period 1982–3 (see also Hanf 1989; Liefferink and Mol 1997). Internalisation policies can indeed have a deregulatory effect, in the sense of favouring a certain room to manoeuvre for private actors over detailed regulation. However, reducing the regulatory burden on industry was only one of the goals of the target-group approach, playing its part in the much broader objective of bringing about a fundamental change in the relationship between economy and ecology. A more equal relationship between public and private actors, a more communicative role of the state, and an emphasis on win–win solutions combining economic and environmental benefits should all contribute to a process towards a full 'embeddedness in our culture' (Winsemius 1986: 61) of environmental concern.

Environmental covenants in practice

The first large-scale experiment based on the target-group philosophy was launched in 1986. It entailed the development by the government and relevant industrial sectors together of measures to reduce emissions of volatile organic compounds by 50 per cent, the so-called KWS-2000 project (see Van Vliet 1992). From around 1990 the target-group approach was systematically put into practice.

In a first phase, the Ministry of Housing, Spatial Planning and the Environment (Ministerie van Volkshuisvesting, Ruimtelijke Ordening en Milieubeheer – VROM) selected fifteen major industrial branches, covering about 90 per cent of total industrial pollution in the Netherlands. On the basis of broad environmental objectives, as laid down, for instance, in the National Environmental Policy Plan (NMP 1988–9), long-term targets for emission reduction were established for each of them. Sectors that were not selected continued to be regulated with the help of traditional licenses.

During the first phase preliminary talks were held with the target groups, but formal consultations with the target groups started only afterwards. With the exception of the long-term reductions targets,[12] all aspects of the measures were open for discussion, including time paths, methods of implementation and monitoring. Apart from the Ministry of VROM – and depending on the sector involved – other ministries (e.g. Economic Affairs, Transport and Public Works) took part in the negotiations. Provincial and municipal authorities were involved in their capacity as licensing authorities under most environmental laws. Industry was represented by the relevant branch organisation, in many cases more of them actually representing different subsectors. Environmental organisations were sometimes consulted in the earlier stages of the process. Negotiations were concluded with an agreement or covenant at the sector level.

In homogeneous sectors, characterised by the application of a limited range of processes and technologies, implementation plans were then worked out at the branch level. They specified, for instance, measures to be taken within a certain time limit, and organisational and procedural arrangements for the communication of the commitments made to the individual companies, and for monitoring and enforcement. Such plans could again take the form of a covenant under civil law.

In heterogeneous sectors implementation plans were elaborated at the level of individual firms. A plan was drawn up by the firm itself in close collaboration with the relevant licensing authority and forms the basis for issuing permits. Due to this link with the traditional procedure, the licensing authorities still have the first responsibility for enforcement at the firm level.

As was pointed out above, covenants have become extremely popular in Dutch environmental policy. The considerable advantages which may account for this include flexibility and speed, particularly in comparison with the lengthy process of the enactment of a formal law. Furthermore, the direct participation of polluters in policy formation is widely claimed, not least by industry itself, to raise their commitment to environmental policy goals. The approach would thus indeed contribute to the internalisation of environmental responsibility. An important advantage for the target groups is, finally, that covenants reduce uncertainty about the future course of government policies. Even though public authorities in principle retain the freedom to 'overrule' covenants by public law, the agreements make drastic changes in the character or content of policies very unlikely (but see Chapter 1). To what extent these advantages actually materialise in practice is difficult to quantify, but it is telling that industry strongly supports the target-group approach. According to the Dutch employers' associations, 'the combination of target group consultation and voluntary action is a better way to bring companies into line with new environmental requirements than simply imposing new measures' (VNO and NCW 1995: 7–8).

The most important shortcomings of covenants in comparison with command and control instruments are, not surprisingly, in the field of implementation and enforcement (see Chapter 8). Some of the earlier environmental covenants, in particular, were formulated as 'gentlemen's agreements', which means that they are difficult to enforce through the courts. Others have a more binding character, but then uncertainty exists as to the consequences of this in practice. As negotiations with industry are usually carried out with intermediary organisations, individual members of such organisations are not legally bound by the outcomes of such negotiations. In some sectors, moreover, the degree of organisation is low, so that parts of the sector are not represented at all. Not surprisingly, problems with regard to the enforcement of covenants are strongly stressed by the environmental organisations and are a reason for them to reject the instrument if it is not used in combination with legislative measures (Biekart 1995).

Enforcement is a problem not only from the point of view of policy effectiveness, however. Also, the sector itself has an interest in preventing free-riding behaviour by firms which do not feel committed to the covenant. For that reason, branch organisations have sometimes themselves suggested the 'affirmation' of a negotiated agreement with the help of formal law. This points to the somewhat paradoxical situation that the most effective way for public authorities to enforce compliance with a covenant – which is, in the end, an agreement under *civil* law – is with the help of *public* law, namely by resorting to customary legislative instruments (more extensively on these points, see Van Vliet 1992; Van Acht 1993; Van Buuren 1993; Van de Peppel 1995).[13]

Another fundamental drawback of covenants and the target-group approach in general is the closed character of the decision-making process. Opportunities for public and political participation and control are limited.

For environmental organisations, this is a second major reason to insist upon combining covenants with formal legislation, established through democratic procedures. Covenants, according to the Netherlands Society for Nature and Environment (Stichting Natuur en Milieu – SNM), are in principle acceptable but 'should be seen primarily as a management instrument for a more effective and efficient implementation of environmental policy targets, always used in combination with legislation and other instruments' (Biekart 1995: 148).

In late 1995 a broad, evaluative study on the use of covenants was published by the Dutch Court of Audit (Algemene Rekenkamer 1995). It covered all 154 covenants that had been concluded between the central government and private actors before 1 September 1994. No less than 85 were concerned with environmental issues. Almost half of those (42) were related to energy saving and had been concluded by the Ministry of Economic Affairs (discussed above). The Ministry of VROM was responsible for another 32. The conclusions of the report confirmed some of the drawbacks of covenants. More than half of all covenants, for instance, turned out to contain insufficient guarantees for effective enforcement. Many of them restricted themselves to mutual commitments to undertake certain efforts, without concrete, measurable goals and deadlines. This figure also applied to the environmental covenants concluded by the Ministry of VROM, but it should be noted that this ministry in particular showed a notable improvement on this point after 1990. The report also confirmed the existence of problems with regard to the actual binding of the target group. In several cases, first, it appeared to be unclear to what extent intermediary organisations covered the entire sector; second, 'hard' guarantees regarding the commitment of individual firms to the agreements made by their representatives were lacking. Finally, the Court of Audit criticised the way covenants were selected as the most appropriate policy instrument. Ministries were often hardly able to substantiate why they expected a covenant to be more effective or efficient in a given context, and whether this outweighed the drawbacks of the instrument.

Region-oriented environmental policy

The regional approach in Dutch environmental policy evolved from the effect-oriented 'track' introduced in the mid-1980s (see above). Until that time, environmental policies with a strong spatial component, for instance the protection of soils or sources of drinking water, had been dealt with quite separately. The first attempts to achieve a higher degree of integration between those 'sectoral' policies, as well as with other, related policy fields, were limited to some areas worth special protection in view of particular ecological values, such as the North Sea and the Wadden Sea. Gradually, however, the broader potential of a regional approach was realised and in

1989 it was presented as a fully fledged third 'track' for the integration of environmental policy. This was particularly apparent in the Fourth Memorandum on Physical Planning (*Vierde Nota over de Ruimtelijke Ordening* 1988–9) and the First NMP (NMP 1988–9), both published under the primary responsibility of the Ministry of VROM. The principal focus of the new policy line, which was also based on negotiation and consensus, was the spatial context of environmental problems, not only in relatively 'clean' areas, but also in problematic regions where various conflicting claims on environmental and spatial resources had resulted in poor environmental quality.

The objective of 'external' integration in the spatial dimension particularly referred to three existing policy fields: physical planning, water policy, and the protection of nature and landscapes (Driessen and Glasbergen 1994). Physical planning has a long tradition in the Netherlands. Characteristic of this policy area is the comparatively high level of decentralisation. Although plans regarding the use of the limited available space are drawn up at all levels of government, the binding municipal 'designation' plans (*bestemmingsplannen*) form the core of the Dutch physical planning system (Brussaard *et al.* 1993). Policies relating to both water quantity and water quality, as we have seen, largely come under the competence of the regional water boards. Protection of nature and landscapes, finally, evidently has a strong spatial component. In 1982 nature conservation was shifted to the present Ministry of Agriculture, Nature Conservation and Fisheries (Landbouw, Natuurbeheer en Visserij – LNV). Together with its competences in the field of agricultural policy, this shift gave the Ministry of LNV a decisive stake in the management of rural areas in the Netherlands.

The primary aim of the Ministry of VROM in developing the third, region-oriented policy line was, of course, to bring about more clearly the environmental aspects of the policy fields of physical planning, water and nature. As they had functioned quite separately, there was indeed a need for integration here. In order to be able to take into account specific demands and conditions, the regional level indeed seemed most appropriate for this purpose. In addition to this, Van Tatenhove (1993) suggested an implicit strategic goal on the part of the Ministry of VROM. Through the process of policy development at the regional level, he argued, the ministry could effectively 'penetrate' issue areas so far dominated by other ministries, notably LNV. The region-oriented approach can thus also be interpreted as a resource in the continuous institutional struggle between VROM and other governmental actors about the definition of the environmental policy domain (Frouws and Van Tatenhove 1993; Van Tatenhove 1996).

The way the region-oriented approach was put into practice is, in many senses, reminiscent of the target-group approach. Five regions with relatively high and six with low environmental quality were selected. In some cases local initiatives were taken up; in other cases the process was instigated entirely by the state level, in particular the Ministry of VROM. In each region, extensive discussion sessions were held with a wide range of actors. Usually, several ministries were involved, as well as provincial and local authorities and water boards. Apart from that, various private interests were represented, such as farmers, industry, recreation and environmental organisations. Much emphasis was put on the development of a common perception of the problems at stake and the creation of a broad basis for future measures. The process was to result in an action plan (*Plan van Aanpak*), signed by all actors but without a legal status. Commitments had to be elaborated in a process of feedback to existing, more sectorised plans and procedures. Most important in this respect are the binding plans that form the core of physical planning policy, but also, for instance, plans regarding water and land management or environmental impact assessment (EIA) can be relevant. In addition, various administrative and financial arrangements, in some cases including considerable subsidies, between central and regional or local government or between public and private actors were often necessary to deal with specific aspects of the action plans.

A detailed evaluation of the region-oriented approach published in 1993 (Glasbergen and Driessen 1993) showed that most of the projects had been successful, in the sense that they had produced (or were at that moment close to finalising) an action plan meeting, to a large extent, the general targets of the approach. In most regions, the process had demonstrably led to a more encompassing and more broadly shared definition of the problems at stake, and sometimes also to a breakthrough in long-existing deadlocks. In several cases, innovative links between environmental and spatial problems were established and issues that had so far been neglected were taken up. Advantages of the region-oriented approach, in short, were related to a more adequate definition of problems and possible solutions, and the establishment of more cooperative relations between the various types of interests involved, rather than in concrete measures. These were to be worked out in a later stage.

The latter observation in fact draws attention to one of the major problems associated with the region-oriented approach. On the basis of the evaluation report, as well as some other literature (Van Tatenhove 1996), these problems can be summarised under the headings *legitimation* and *operationalisation*. In the first place, a serious bottleneck in the first phase of the process turned out to be the selection of the actors to be involved in the negotiations. Too large a group would make the process unworkable, but a too limited selection would jeopardise the legitimation basis of the outcomes. Although various interest groups, including environmental organisations, were in most cases represented in the negotiations, support for (and even knowledge of) the process among the general population was sometimes limited. In addition, as traditional democratic institutions hardly play a role in the process itself, Glasbergen and Driessen (1993: 142) argue for more formalised arrangements for public participation.

Second, the operationalisation of the mostly global agreements laid down in the action plan carries with it considerable uncertainties, due both to the non-binding character of the plan and to problems that may arise later. An important reason for such problems is that the region-oriented approach is laid 'on top' of existing policies rather than replacing them. The necessary feedback to existing instruments and procedures in the operationalisation phase often brings older command and control measures back into play. Taking into account regional conditions – which is in fact one of the essential assets of the region-oriented approach as such – may, for instance, be difficult if concrete plans or measures have to be worked out and related to generic, nationwide laws, standards or policy objectives. It may also happen that procedures that have to be followed in a later stage, as it were, unsettle the consensus reached earlier. This can, for instance, be the case in the procedure for IEA, where there is a legal obligation to consider a number of alternative options.

Despite the problems just sketched, the region-oriented approach has a decentralising effect which may, in turn, affect existing political and institutional relations in the policy fields involved.[14] As far as rural areas are involved, Frouws and Van Tatenhove (1993) point out that the region-oriented approach adds both to the forces undermining the highly centralised system of agrarian corporatism in the Netherlands and to the tendency of a *rapprochement* between the environmental and agricultural policy fields. In the rural area of the *Gelderse Vallei* (Gelderland Valley), the Ministries of VROM and LNV, whose views had formerly often been diametrically opposed, were found to develop more and more mutually shared models for the development of the region. As a result, they alienated themselves, to some extent, from their respective constituencies, the environmental movement and the farmers' organisations, who now occupied the more extreme positions (see Van Tatenhove 1993). The region-oriented approach thus had the effect of shifting the conflict from within the state to a broader societal context, with the government increasingly in the role of facilitator and mediator. In this sense, there is an obvious parallel with the target-group approach, putting a similar emphasis on negotiation and consensus beyond the limits of state institutions, on the centrality of the process rather than the outcomes, and on the creation of win–win situations.[15] The latter aspect seems to be particularly important for the region-oriented approach. The construction of package deals from which all parties can gain in one way or another can be seen as the very basis of the commitment of the entire range of extremely multifarious actors in the process (Ten Heuvelhof and Termeer 1991).

Conclusions

This chapter shows that the phenomenon of 'new' instruments in environmental policy cannot be considered in isolation. Instead, to understand their emergence and impact one must view them in relation to the broader policy context in which they are applied. In the Netherlands a decisive shift in the approach to environmental policy-making took place in the mid-1980s. The leading ideas behind this shift were the concepts of integration and the internalisation of environmental responsibility throughout society. It led, among other things, to the development of the target-group and region-oriented approaches. The common denominator of the new approach was the focus on communication and reflection about both the form and the content of environmental policy and the involvement of both public and private actors in these processes. In this sense, both the target-group approach and the region-oriented approach can be associated with the concept of ecological modernisation.

Apart from a changing, more 'stage-setting' role for the state, the theory of ecological modernisation also stipulates an increasing importance of market dynamics in ecological restructuring. Although economic instruments can certainly be seen in that context as well, and are indeed the dominant expression of ecological modernisation in some other countries, they have not been particularly important in the Netherlands so far. The only notable exceptions to this are the system of effluent charges on water pollution and a number of financial arrangements mainly aimed at consumers, including the recently established energy tax for small users. The effluent charge, however, was already introduced long before the ideological reform of Dutch environmental policy and can therefore hardly be associated with it. Although its significant regulative effect was recognised, particularly in the 1980s, the instrument did not find wider application for industrial sources. This may be related to the fact that, as argued earlier, the introduction and success of the effluent charge was at least partly dependent on a specific institutional constellation.

The shift away from a traditional command and control approach in Dutch environmental policy thus focused on negotiation and consensus, partnership and shared responsibility, and the search for synergy between economy and ecology. This approach now dominates industrial environmental policy and has come to play an important role in regional policy as well. In practice, the communicative approach relies on a broad range of policy instruments in a more narrow sense of the word. Implementation of covenants and action plans may involve relatively unorthodox elements such as the development of environmental management systems, ecoaudits, but there are most often also links with 'old' mandatory standards, licensing systems and planning procedures. The difference with the traditional command and control system is, rather, in the increased openness of options and the establishment of a continuous dialogue between public and private actors on fundamental policy choices.

Notes

1 I would like to thank Arthur Mol, Jan van Tatenhove, Jaap Frouws and the editor of this volume for their help in preparing this chapter and for valuable comments on earlier versions.
2 See, for instance, Spaargaren and Mol 1992; Mol and Spaargaren 1993; Hajer 1995; Mol 1995; Hannigan 1995. On reflexive modernisation generally, see Giddens 1990; Beck 1992; Beck *et al.* 1994.
3 For a general discussion of Dutch environmental policy, covering both history and organisational features, see Liefferink (1997).
4 The administrative structure of Dutch water-quality management is, in fact, quite complicated. For an adequate and concise description in English, see Andersen (1994: ch. 7).
5 In the field of waste collection it should be noted that separation of household waste into organic and non-organic fractions, glass, paper, etc. is now practised in various forms practically everywhere, but this is not accompanied by economic incentives. Some small-scale experiments have shown that there are still considerable technical barriers to 'pricing' directly the amount of waste produced by each household.
6 This holds as long as consumer prices are not too much different from those in neighbouring countries. In the past it has been demonstrated that, for instance, high petrol prices may entice significant numbers of people in border regions to shop abroad.
7 This kind of problem, to be sure, was not unique for the Netherlands and can be linked to more general discussions on 'state failure' in, among others, environmental policy in this period (Jänicke 1986).
8 In an international context, Winsemius strongly propagated systems for the compensation of unevenly distributed costs of anti-pollution measures, for instance in the field of acidification (Winsemius 1986: 132–3).
9 Environmental covenants are sometimes labelled 'voluntary agreements', but this term does not appear to be fully appropriate. In the first place, as explained in this section, covenants are to be seen as part of a broader policy approach in which 'harder' policy instruments (general legal obligations, licences) may play a role as well. In the second place, even if this context is not taken into account, the state in its capacity of public actor may still use the introduction of formal regulation as a stick with which to beat industry if 'voluntary' negotiations do not bring the desired results (Liefferink and Mol 1997).
10 For this reason, the target-group approach should be associated with meso- rather than with macro-corporatism. The notion of meso-corporatism was developed in the 1980s to deal with corporatist arrangements at the sectoral level, usually involving only the state and business associations (see particularly Cawson 1985).
11 The relationship between the Dutch target-group approach and corporatism is one of the subjects of an ongoing comparative research project on the politics of negotiated agreements in the environmental field by the author and Arthur Mol in collaboration with the Universities of Aarhus and Salzburg.
12 However, Mol points out that in the early case of the KWS-2000 project, industry representatives presumed that also the 50 per cent reduction goal itself was negotiable (Mol 1995: 154).

13 A small part of the covenants in Dutch environmental policy were, from the beginning, intended to be superseded by formal legislation (see Algemene Rekenkamer 1995: 13–14). In that case, the problems of enforcement addressed here obviously do not exist.
14 It might even be argued that the occurrence of conflicts between agreements at the regional level and central plans and procedures illustrates the success and the effective decentralising impact of the region-oriented approach.
15 It should be borne in mind, however, that policy coordination through negotiation and planning has a considerably longer history in the field of physical planning than in the environmental field. However, the involvement of various kinds of private actors in this process is new also for physical planning (Van Tatenhove and Van den Aarsen 1996).

References

Algemene Rekenkamer (1995) *Convenanten van het Rijk met Bedrijven en Instellingen*, Tweede Kamer 1995–6, 24 480, nos. 1–2

Andersen, M. S. (1994) *Governance by Green Taxes: Making Pollution Prevention Pay* (Manchester: Manchester University Press).

Beck, U. (1992) *Risk Society: Towards a New Modernity* (Oxford: Blackwell).

Beck, U., Giddens, A. and Lash, S. (1994) *Reflexive Modernization: Politics, Tradition and Aesthetics in the Modern Social Order* (Oxford: Polity Press).

Bennett, G. (ed.) (1991) *Air Pollution Control in the European Community. Implementation of the EC Directives in the Twelve Member States* (London: Graham & Trotman).

Biekart, J. W. (1995) 'Environmental covenants between government and industry: a Dutch NGO's experience', *Reciel* 4(2): 141–9.

Bressers, H. (1983a) 'The role of effluent charges in Dutch water quality policy', in P. B. Downing and K. Hanf (eds) *International Comparisons in Implementing Pollution Laws* (Boston, MA: Kluwer-Nijhoff Publishing).

Bressers, J. T. A. (1983b) *Beleidseffektiviteit en Waterkwaliteitsbeleid. Een Bestuurskundig Onderzoek* (Enschede: Technische Hogeschool Twente), dissertation.

—— (1995) 'The impact of effluent charges: a Dutch success story', in M. Jänicke and H. Weidner (eds) *Successful Environmental Policy: A Critical Evaluation of 24 Cases* (Berlin: Edition Sigma).

Bressers, J. T. A. and Schuddeboom, J. (1994) 'A survey of effluent charges and other economic instruments in Dutch environmental policy', in OECD *Applying Economic Instruments to Environmental Policies in OECD and Dynamic Non-member Economies* (Paris: OECD).

Brussaard, W., Drupsteen, T. G., Gilhuis, P. C. and Koeman, N. S. J. (eds) (1993) *Milieurecht*, 3rd edn (Zwolle: W. E. J. Tjeenk Willink).

Cawson, A. (ed.) (1985) *Organised Interests and the State: Studies in Meso-corporatism* (London: Sage).

—— (1986) *Corporatism and Political Theory* (Oxford: Blackwell).

Driessen, P. P. J. and Glasbergen, P. (1994) 'Het Gebiedsgericht Milieubeleid', in P. Glasbergen (ed.) *Milieubeleid, een Beleidswetenschappelijke Inleiding*, 4th edn (Den Haag: VUGA).

Frouws, J. (1993) *Mest en Macht. Een Politiek-Sociologische Studie naar Belangenbehartiging en Beleidsvorming inzake de Mestproblematiek in Nederland vanaf 1970* (Wageningen: WAU), dissertation.

Frouws, J. and Van Tatenhove, J. (1993) 'Agriculture, environment and the state: the development of agro-environmental policy-making in the Netherlands', *Sociologia Ruralis* 33(2): 220–39.

Giddens, A. (1990) *The Consequences of Modernity* (Oxford: Polity Press).

Glasbergen, P. and Driessen, P. P. J. (eds) (1993) *Innovatie in het Gebiedsgericht Beleid. Analyse en Beoordeling van het ROM-gebiedenbeleid* ('s Gravenhage: SDU Uitgeverij Plantijnstraat).

Hajer, M. A. (1995) *The Politics of Environmental Discourse: Ecological Modernization and the Policy Process* (Oxford: Clarendon Press).

Hanf, K. (1989) 'Deregulation as regulatory reform: the case of environmental policy in the Netherlands', *European Journal of Political Research* 17: 193–207.

Hannigan, J. A. (1995) *Environmental Sociology: A Social Constructionist Perspective* (London: Routledge).

Huber, J. (1982) *Die verlorene Unschuld der Ökologie: neue Technologien und superindustrielle Entwicklung* (Frankfurt am Main: Fischer Verlag).

IMP-M (*Indicatief Meerjaren Programma Milieubeheer*) (1985–9) Tweede Kamer 1984–1985, 18602, nos. 1–2.

—— (1986–90) Tweede Kamer 1985–1986, 19204, nos. 1–2.

Jaarsma, E. and Mol, A. P. J. (1994) 'De Rol van het Onderzoek in het Beleidsproces rond Regulerende Energieheffingen', *Milieu* 9(3): 120–8.

Jänicke, M. (1986) *Staatsversagen: die Ohnmacht der Politik in der Industriegesellschaft* (Munich: Piper).

—— (1993) 'Über ökologische und politische Modernisierungen', *Zeitschrift für Umweltpolitik und Umweltrecht* 2: 159–75.

Le Blansch, K. (1996) *Milieuzorg in Bedrijven. Overheidssturing in het Perspectief van de Verinnerlijkingsbeleidslijn* (Amsterdam: Thesis Publishers).

Leroy, P. (1994) 'De Ontwikkeling van het Milieubeleid en de Milieubeleidstheorie', in P. Glasbergen (ed.) *Milieubeleid, een Beleidswetenschappelijke Inleiding*, 4th edn (Den Haag: VUGA).

Liefferink, D. (1996) *Environment and the Nation-state: The Netherlands, the European Union and Acid Rain* (Manchester: Manchester University Press).

—— (1997) 'The Netherlands: a net exporter of environmental policy concepts', in M. S. Andersen and D. Liefferink (eds) *European Environmental Policy: The Pioneers* (Manchester: Manchester University Press).

Liefferink, D. and Mol, A. P. J. (1997) 'Voluntary agreements as a form of deregulation? The Dutch experience', in U. Collier (ed.) *Deregulation in the European Union: Environmental Perspectives* (London: Routledge).

Mol, A. P. J. (1995) *The Refinement of Production: Ecological Modernisation Theory and the Chemical Industry* (Utrecht: Van Arkel).

Mol, A. P. J. and Spaargaren, G. (1993) 'Environment, modernity and the risk society: the apocalyptic horizon of environmental reform', *International Sociology* 8(4): 431–59.

NMP (*Nationaal Milieubeleidsplan*) (1988–9) Tweede Kamer 1988–1989, 21137, nos. 1–2.

Opschoor, J. B. (1994) 'Economische Politiek, Milieubeleid en Beleidsinstrumenten', in F. Dietz, W. Hafkamp and J. van der Straaten (eds) *Basisboek Milieu-economie* (Amsterdam: Boom).

Schuurman, J. (1988) *De Prijs van Water. Een Onderzoek naar Aard en Omvang van de Regulerende Nevenwerking van de Verontreinigingsheffing Oppervlaktewateren* (Arnhem: Gouda Quint).

Spaargaren, G. and Mol, A. P. J. (1992) 'Sociology, Environment and Modernity: Ecological Modernization as a Theory of Social Change', *Society and Natural Resources* 5: 323–44.

Steering Committee on Regulating Energy Taxes (1992) *Report of the Independent Research into the Administrative Possibilities, as Well as the Energy and Economic Impacts of the Introduction of Regulating Energy Taxes* ('s Gravenhage: Steering Committee on Regulating Energy Taxes).

Ten Heuvelhof, E. and Termeer, K. (1991) 'Gebiedsgericht Beleid en het Bereiken van Win-Win-Situaties', *MBGB* 4: 8–9.

Van Acht, R. J. J. (1993) 'Afdwingbare Milieuconvenanten?', *Nederlands Juristenblad* 67(14): 512–17.

Van Buuren, P. J. J. (1993) 'Environmental covenants – possibilities and impossibilities: an administrative lawyer's view', in J. M. van Dunné (ed.) *Environmental Contracts and Covenants: New Instruments for a Realistic Environmental Policy?* (Lelystad: Koninklijke Vermande).

Van de Peppel, R. A. (1995) *Naleving van Milieurecht. Toepassing van Beleidsinstrumenten op de Nederlandse Verfindustrie* (Deventer: Kluwer).

Van Putten, J. (1982) 'Policy style in the Netherlands: negotiation and conflict', in J. Richardson (ed.) *Policy Styles in Western Europe* (London: George Allen & Unwin).

Van Tatenhove, J. (1993) *Milieubeleid Onder Dak Beleidsvoeringsprocessen in het Nederlandse Milieubeleid in de Periode 1970–1990: Nader Uitgewerkt voor de Gelderse Vallei* (Wageningen: Pudoc), Wageningse Sociologische Studies no. 35.

—— (1996) 'De Regio als Beleidsarena', *Sociologische Gids* 1: 46–59.

Van Tatenhove, J. P. M. and Liefferink, J. D. (1992) 'Environmental policy in the Netherlands and in the European Community: a conceptual approach', in F. von Benda-Beckmann and M. van der Velde (eds) *Law as a Resource in Agrarian Struggles* (Wageningen: Pudoc), Wageningen Studies in Sociology no. 33.

Van Tatenhove, J. and Van den Aarsen, L. (1996) 'Politieke Modernisering en Doelgroepenbeleid voor het Landelijk Gebied', *Tijdschrift voor Socialwafenschappelijk Onderzoek van de Landbouw* 11(4): 253–74.

Van Vliet, L. M. (1992) *Communicatieve Besturing van het Milieuhandelen van Ondernemingen: Mogelijkheden en Beperkingen* (Delft: Eburon).

Vermeulen, W. J. V. (1994) 'Het Economische Sturingsmodel', in P. Glasbergen (ed.) *Milieubeleid. Een Beleidswetenschappelijke Inleiding*, 4th edn ('s Gravenhage: VUGA).

Vierde Nota over de Ruimtelijke Ordening (1988–9) Tweede Kamer 1988–1989, 20490, nos. 9–10.

VNO and NCW (1985) *Het Milieubeleid Nader Bekeken* ('s Gravenhage: Verbond van Nederlandse Ondernemingen/Nederlands Christelijk Werkgeversverbond).

—— (1995) *Environmental Policy in the Netherlands: The Role of Industry* ('s Gravenhage: Verbond van Nederlandse Ondernemingen/Nederlands Christelijk Werkge-Versverbond).

Wallace, D. (1995) *Environmental Policy and Industrial Innovation: Strategies in Europe, the US and Japan* (London: Earthscan/Royal Institute of International Affairs).
Weale, A. (1992) *The New Politics of Pollution* (Manchester: Manchester University Press).
WRR (Wetenschappelijke Raad voor het Regeringsbeleid) (1992) *Milieubeleid. Strategie, Instrumenten en Handhaafbaarheid* ('s Gravenhage: SDU), WRR Rapporten aan de Regering no. 41.

5

NEW ENVIRONMENTAL POLICY INSTRUMENTS IN BELGIUM

Kurt Deketelaere

Introduction

In Belgium, as in other arenas and levels of policy-making, a gradual change is taking place in the choice of environmental policy instruments: where environmental policy until recently was exclusively characterised by command and control, the market-based approach is becoming increasingly popular (Gaines and Westin 1991; OECD 1993; K. Deketelaere and Martens 1994; K. Deketelaere and Pittevils 1995).

This contribution will briefly analyse the Belgian experience with traditional instruments of environmental protection and then explore the ways in which a variety of new environmental policy instruments have been applied. Belgium is a federal state, made up of three communities and three regions. The latter, which include the Flemish Region, the Walloon Region and the Brussels-Capital Region, have substantial environmental competences. As a consequence, five sets of environmental legislation must be taken into account: Flemish, Walloon, Brussels, federal and European environmental legislation. While each of these will be discussed at various points throughout this chapter, the analysis will primarily deal with Flemish environmental legislation and federal environmental legislation which is applicable in the Flemish Region.

Environmental protection in Belgium

In many European countries, including Belgium, the use of instruments of direct regulation (prohibitions, restrictions, permit systems, notification systems) in environmental policy prevailed in the 1960s, 1970s, 1980s and even the (early) 1990s (OECD 1993). The advantages of these instruments, known as command and control, are well known: they establish clear environmental norms which must be met, these norms apply to everybody, the government must not evaluate the individual circumstances of thousands of different cases, and the use of general norms limits administrative discretionary power and makes it easier for companies to plan their own environmental policy.

On the basis of these classic arguments, Belgium traditionally made enormous use of command and control environmental regulation. Federal environmental legislation – town and country planning (1962), air-pollution control (1965), protection of surface waters (1971), protection against noise (1973), nature protection (1973), the management of risks of heavy accidents with certain industrial activities (1987), and protection against ionising radiations (1994) – and also (later) regional (Flemish) environmental legislation – waste management (1981 and 1994), groundwater management (1984), environmental permits (1985, 1991 and 1995), environmental impact assessment (EIA) (1989), protection of forests (1990), manure (1991), protection of dunes (1993), management of gravel extraction (1993), environmental policy agreements (1994), environmental planning (1995), environmental care at plant level (1995), soil sanitation (1995), spatial planning (1996) and nature conservation (1997) – established a wide variety of prohibitions and restrictions. It was not only the original versions of these laws and decrees that contained those instruments of direct regulations. Later modifications and updates also used these instruments, albeit not (more) exclusively but in combination with other environmental policy instruments.

As an example of this command and control approach, which is still current, the Flemish government fixed in 1995, for the first time, a general legal basis for the adoption and use of environmental-quality norms for the protection of the environment (although such quality norms had already been used for a number of years in the framework of specific sectoral laws and decrees) (Lefebure 1996). These norms indicate the maximum allowable levels of pollutants in the atmosphere, the water or the soil. They can also determine which natural or other elements must be present in the environment in view of the protection of the ecosystems and the promotion of biological diversity. A distinction must be made between basic environmental-quality norms (which establish quality demands throughout the whole Flemish Region) and specific environmental-quality norms (which apply in areas which need special protection). Environmental-quality norms can be fixed in the form of limit values (which may not be exceeded) and directional values (which must be achieved as much as possible or maintained).

Environmental-quality norms (as fixed in Part 2 of the so-called VLAREM II, the decision of the Flemish government of 1 June 1995) are used in the Flemish Region for noise, surface water, soil, groundwater and air (Lefebure 1996). Sustainable development and the protection of a healthy environment are also considered general basic environmental-quality norms.

The general and sectoral environmental conditions for classified installations, as embedded in Part 4 and Part 5 of VLAREM II, are considered to conform with the best available technologies (BAT) (Gille and Lambrecht 1996). VLAREM II also prescribes that every licensee must apply BAT in order to protect people and the environment. Before the adoption of VLAREM II (original version 1992, current version 1995) there was no general obligation to apply BAT. In that period, application of BAT was only foreseen in laws and/or decrees implementing European environmental Directives which imposed the application of BAT. This means that environmental legislation, and the instruments of direct regulation in that legislation, in that period did not always reflect the BAT of that time. Rather, most instruments of direct regulation arose from political and technical consultation groups, composed of representatives of all concerned parties (government, industry, green non-governmental organisations (NGOs), etc.). Real and clear interference by political parties in this process of standard-setting was and is still exceptional. A clear example in Flemish environmental policy, however, was the adoption of the 'manure action plan' (De Batselier 1996): because of the alliance of the agricultural organisations with the Christian Democratic Party, this party intervened explicitly in the setting of standards for the production and spreading of manure; this led to a real confrontation with the former Socialist minister for the environment, Norbert De Batselier, regardless of the fact that the Socialist Party was at that time the coalition partner of the Christian Democratic Party. Because almost no farmers vote for the Socialist Party, this party could go very far in (proposals for) the setting of manure standards and in this way also portray itself as a 'green' party.

The introduction of new environmental policy instruments

During the 1980s and the early 1990s awareness grew that environmental policy goals could no longer be achieved by instruments of direct regulation alone. The main disadvantages of instruments of direct regulation are that they are often inflexible and economically inefficient, and that they discourage the development of clean technology (OECD 1993: ch. 1).

This new awareness had different causes. First of all, there was increased pressure from public opinion. Several environmental catastrophes (Bhopal, Seveso, Amoco Cadiz) and the publication of several (alarming) international, European, national and regional environmental reports increased substantially the environmental awareness of public opinion, which demanded more stringent and efficient environmental policy. Second, this increased environmental awareness was translated in a more concrete and active way by the green NGOs, who enjoyed great support during the 1980s and the early 1990s. They managed to put the problem of the efficiency of environmental policy on the political agenda and to force the government in a more or less new direction. Third, at the level of the Flemish government there was an ambitious Christian-Democratic minister for the environment, who, for the first time, introduced a long-term policy vision concerning real and effective environmental protection, with ambitious goals. Fourth – and this is, in the author's opinion, the most important reason for a shift in the choice of environmental policy instruments – there was a lack of (Flemish) government means (both technical and financial). This led primarily to the introduction of (financing) environmental taxes (lack of financial means), environmental policy agreements and environmental care systems (lack of enforcement means).

Fifth, the position of industry has always been a bit ambiguous. On the one hand, it was not in fact demanding a change of environmental policy (instruments). This is quite understandable because instruments of direct regulation are the kind of instruments industry can best cope with. However, because they were often too stringent at short notice and changed very often, Flemish instruments of direct regulation have been heavily contested by industry in recent years. On the other hand, industry has always had a positive and cooperative reaction to new environmental policy instruments on the condition that they were drafted and introduced in what it deemed to be a reasonable way. Because this was not always the case, on the regional or federal level, industry has also often opposed new environmental policy instruments. Finally, interest in expanding the arsenal of federal and regional environmental policy instruments was also highly influenced by the declarations, initiatives and activities of the EC and the OECD in this field (see Chapter 1).

In light of these reasons or appeals for reform, the number of Belgian instruments of direct regulation was supplemented by a variety of new instruments devoted to market regulation, social regulation, planning and financial aid (K. Deketelaere 1991, 1992, 1995; K. Deketelaere and Martens 1994, 1996; K. Deketelaere and Pittevils 1995). So, this means that, in fact, (1) the number of instruments of direct regulation did not decline and (2) a lot of those instruments were not replaced by other instruments.

All of this happened in a period of consolidation of the environmental competences of the regions (by the third (1988) and fourth (1993) reform of the state), as well as in the period of drafting the first Environment and Nature Plan for the Flemish Region (1990). But perhaps the most significant catalyst in this policy period, and in this policy shift towards new instruments, was the formation by the (former and current) Flemish minister for the environment, Theo Kelchtermans, of the Interuniversity Commission for the Reform of Environmental Legislation in the Flemish Region, which operated during the period 1989–95 (IUCHM 1995). The task of that commission was to integrate environmental concerns into other policy areas, but at the same time to rationalise and simplify environmental legislation, increase its efficiency and improve means for its enforcement. The Interuniversity

Commission's draft framework decree on environmental policy led to several of the new instruments discussed below. In the future more new decrees will be adopted and more existing environmental legislation will be modified on the basis of this draft.

The creation of the Interuniversity Commission was not a consequence of pressure from industry or green NGOs. It was the consequence of a combined political–academic initiative. On the one hand, as mentioned before, a new ambitious Christian-Democratic minister for the environment was nominated; the Department of the Environment became more and more important, not only for reasons of public opinion but also for reasons of budget figures and political power. On the other hand, there was the academic world, which was strongly interested, for academic and financial reasons, in the enormous project of rewriting Flemish environmental legislation.

However, somewhat in contrast with his academic-legislative initiative, this same minister for the environment also produced a lot of environmental legislation during his first term (1989–92) (K. Deketelaere 1994c). A great percentage of the legislation of this period can be characterised as command and control: a large number of new prohibitions, restrictions and quality norms were introduced. However, it is also in this period that a start was made with the introduction of new environmental policy instruments. For example, for financial reasons, financing environmental taxes were made operational from the early 1990s.

Because in recent years the instruments of direct regulation were only supplemented with other environmental policy instruments and not replaced, the most important reproach to the current minister for the environment, Theo Kelchtermans, is that until now only new (even more stringent) environmental legislation has been created (consisting of all kinds of environmental policy instruments) but no environmental legislation has yet been abolished. In order to meet with this demand, Mr Kelchtermans has created a new commission (Commission for the Evaluation of Environmental Legislation), which, in the first place, must make a technical and economic evaluation of the legislation concerning environmental permit demands and procedures. The intention is that after this specific evaluation a global evaluation of the whole environmental legislation will be undertaken, from a technical, economic and legal perspective.

New instruments in practice

Environmental levies

The recently developed environmental tax legislation in Belgium includes two forms of environmental levies: *financing environmental levies* and *regulatory environmental levies*.

Financing environmental levies

Financing environmental levies (levies aimed at financing the environmental policy of the government, be it totally or partially) were introduced throughout the 1980s and have become the most widely used new environmental instrument of market regulation in Belgian (K. Deketelaere and Pittevils 1995) and Flemish (K. Deketelaere and Martens 1994) environmental policy.

At the federal level, they were introduced for certain industrial activities which can cause heavy accidents (1990), the private use of energy (1993), ionising radiations (1994) and dangerous products (1994), and are deposited in specific sectoral funds which finance a few very expensive policy fields (nuclear plants, Seveso-companies, emergency planning, ionising radiations).

In addition to federal action, the Flemish Region established levies on the removal of waste (1986 and 1990), the pollution of surface water (1990), the overproduction of manure (1991), the delivery of a permit for the intake of water (1990), the extraction of gravel (1993), and the import or export of waste (1994). The Walloon Region has also introduced in recent years levies on the discharge of industrial and household waste water (1990), the removal of waste (1991), and on groundwater and drinking water (1990). The Brussels-Capital Region has adopted the following financing environmental levies: a levy on the delivery of environmental permits (1992) and a levy on the discharge of waste water (1996). The revenues of the most important Flemish environmental levies are deposited in the so-called MINA-Fund and finance, especially, the building of water purification plants, sewerage systems and the regional environmental institutions.

The obvious advantage of financing environmental levies is, of course, that they can raise a lot of money. These levies are often the most important financial source for funding environmental policy of the competent authority. The disadvantage of financing environmental levies is that they are not a correct implementation of the polluter-pays principle. They are in general too low for the large polluters and too high for the small polluters: the taxed polluting products, services or activities do not disappear from the market because the large polluter can easily pay the levy and decide not to change its behaviour; the small polluter can just pay the levy but does not change its behaviour because no money is left to finance cleaner alternative production methods.

Although every financing environmental levy also has a certain regulatory effect, most of the federal and regional environmental levies in Belgium are aimed at revenue-raising rather than providing incentives for environmentally friendly changes in the behaviour of polluters; their contribution to the improvement of the environment is limited and their contribution to the budget funds concerned is significant (Van Humbeeck 1992).

This was proved by different studies of the Flemish Steering and

Working Group on Environmental Levies (1993): the revenues of the most important Flemish environmental levies amount approximately to Bfr10 billion, currently the most important source of financing for Flemish environmental policy. A similar environmental policy without those financial means would not be possible in the Flemish region, certainly not if the only alternative were to draw funding from the general budget of the Flemish government. The same studies have indicated that the environmental benefits created by those financing environmental levies are not impressive. On the contrary, any improvements made in environmental quality in the Flemish Region will certainly not have resulted from the new (financing) environmental levies alone. The recent and more stringent environmental instruments of direct regulation have been at least as important (for similar findings in other states, see Chapters 2, 3 and 4).

It is self-evident that the introduction in recent years of financing environmental levies has been (heavily) contested, by green NGOs as well as by industry. Green NGOs and political parties have continually pleaded for regulatory environmental levies (Steenwegen 1993a, 1993b; Pauwels and Decoster 1995); industry has always asked for environmental levies which are 'society-relevant' (i.e. which take into account the pollution caused not only by industry, but also by households and farmers) and which implement the polluter-pays principle (financing environmental levies violate this with a minimum rate and amount, to be paid by all possible taxpayers) (VEV 1995a, 1995b).

Regulatory environmental levies

In contrast to financing environmental levies, the big advantage of regulating environmental levies is that, in general, they seek to change the incentives facing consumers and producers rather than simply raising tax revenue (OECD 1993). Levies are high enough and targeted well enough that the taxed polluting products, activities or services disappear from the market in a short period of time. These levies are a real implementation of the polluter-pays principle.

However, it must be said that regulatory environmental levies are often characterised by (1) a wrong choice of levy base; (2) a wrong choice of levy rate; (3) bad timing of introduction; (4) poor coordination with other policy areas (which influence environmental policy) (OECD 1993). This will be illustrated below with the Belgian 'ecotax'.

The 1993 levy on products damaging the environment ('ecotax') is currently the only real regulatory environmental levy used in regional and/or federal environmental policy in Belgium (Van Orshoven 1993; K. Deketelaere 1994a, 1994b, 1996a). As concerns the motives for its introduction, they are quite strange. In 1993, when the fourth reform of the state had to be approved by the federal parliament, the ruling majority needed the votes of some opposition parties, because the approval needed a two-thirds majority. Green parties (AGALEV, ECOLO), in particular, were the ones who furnished the necessary votes. In return, however, they demanded an ecotax on packaging, packaging waste and environmentally unfriendly products. In this way, the ecotax was born, quite unexpectedly and without preparation.

The levy concerned the following products: beverage packaging; throw-away cameras and razors; batteries; packaging for certain industrial products; insecticides; paper. In most cases, the levy could be avoided (by meeting recycling and/or re-use percentages or by establishing deposit–refund systems), or reductions and exemptions could be obtained. The levy was equated with excise duties and placed on a product because of the damage which it was deemed to cause to the environment.

However, when the levy on products damaging the environment was introduced in 1993 in Belgium, all of the above-mentioned possible problems with regulatory environmental levies appeared (K. Deketelaere 1994a, 1994b, 1996a) and provoked a lot of protests from industry (VEV 1995b):

1 The choice of levy base was wrong, in that, in some cases, the levy base could not be controlled or there were no environmentally friendly alternatives to the taxed products (this being one of the basic ideas and conditions for the choice of taxed products). This was particularly problematic for the levy on paper (the levy base, being the amount of recycled fibres in the paper, was unverifiable) and insecticides (many insecticides were taxed, although there was no environmentally friendly alternative to all the insecticides concerned).
2 The choice of levy rate was wrong. It was often too high given the absence of an environmentally friendly alternative to the taxed product and the introduction at short notice of the tax. This was particularly problematic for the levy on throw-away cameras (300 BF/camera), throw-away razors (10BF/razor) and batteries (20 BF/battery).
3 The introduction of the levy was poorly timed. For certain products it was due almost immediately. This was a major problem for the levy on beverage packagings (1 January 1994), batteries (1 January 1994), industrial packaging (1 January 1994) and paper (1 January 1994). Given that the levy was only published in the *Belgian State Gazette* of 20 July 1993, it was impossible for industry to adjust its production processes in order to avoid the levy.
4 There was poor coordination with other policy areas. It was introduced as a federal levy and no account was taken of regional environmental policy (particularly regional waste policy, which already foresaw some ecotaxed products), federal fiscal policy (which had its own classification of products subjected to excise duties) or European environmental policy (where product-oriented environmental policy is governed by

notification and technical information obligations (Directive 83/189) as well as by Directives 75/442 on waste, 91/157 on batteries, and 94/62 on packaging and packaging waste).

As a consequence of all these problems, the different levies were repeatedly postponed for most of the ecotaxed products. Often demanded by industry because of economic reasons, postponements were always condemned by green NGOs and political parties (Steenwegen 1993a, 1993b; Pauwels and Decoster 1995). This was, for example, the case with the choice for recycling of beverage packaging instead of re-use, and with the reduction in the number of ecotaxed pesticides.

In its present form, the ecotax can in fact be more accurately characterised as a financing environmental levy than as a regulatory environmental levy. The main lines of the present regulation (originally of 16 July 1993 and frequently modified up through 1997) can be summarised as follows (K. Deketelaere 1996a):

- A Bfr15 levy on all packaging brought into consumption, regardless of the contents, the cubic measure or the material from which it is produced. Exemptions are made for the setting up of a deposit system for reusable packaging, or for attaining specified recycling percentages for recyclable packaging. This levy came into force on the following dates: 1 April 1994 for packaging of beer, soda water, cola and other lemonades; 1 January 1996 for packaging of other beverages.
- A Bfr300 levy on disposable cameras. Exemptions are made for the setting up of a collection system which attains specified re-use and/or recycling percentages. This levy came into force on 1 July 1994.
- A Bfr20 levy on batteries. Exemptions are made for setting up a deposit system, return premium system, or collection and recycling system. This levy came into force on 1 January 1996.
- A Bfr25 minimum levy per volume-unit of packaging (in general 5 litres), with a maximum of Bfr500/packaging. Exemptions are made in the case of setting up a deposit system for gum, solvents and insecticides brought into consumption and which are used professionally. This tax came into force on 1 January 1996.
- A Bfr10 or Bfr2 levy per gram of certain active substances. Some exemptions are made for insecticides. This ecotax came into force on 1 July 1996.
- A Bfr10 levy per kilogram of paper or cardboard unless specified proportions of recycled fibres are attained. A number of (inconsistent) exemptions are forseen. This tax came into effect on 1 January 1998.

As regards financial impact, one can only say that this ecotax, until now, has been an expensive matter for the federal government. Its installation and application has cost a few tens or hundreds of million Belgian francs, while the revenues, although they are not pursued (because it is a regulatory environmental levy), amount so far to only a few million francs, which, moreover, are distributed among the regions for environmental purposes.

As regards efficiency, it must be admitted that the ecotax, in spite of the many problems, has been very successful (K. Deketelaere 1995, 1996a). Several years before the compulsory introduction of measures concerning packaging and packaging waste (in 1996, implementing Directive 94/62/EC), the Belgian ecotax had already forced industry to take measures in this field. With the ecotax as a stick behind the door, different branches of Belgian industry managed to adjust their products and production processes from an environmental point of view. As mentioned before, specific, environmentally friendly collection systems were set up, for example for beverage packaging, batteries and disposable cameras. Of course, this also influenced consumer behaviour.

Instruments of financial aid

Although the federal government has been reducing the (fiscal) possibilities for investment deduction in general, the federal Code of Income Taxes provides, since the beginning of the 1990s, for an increased investment deduction for environmentally friendly investments (K. Deketelaere and Martens 1996). The standard deduction is increased (up to 13.5 per cent in 1996) when the investment concerns the development of new products and future technologies which aim at limiting negative environmental effects as much as possible or which increase industrial energy efficiency.

Increased (regional) expansion aid for environmentally friendly investments is another example of a (fiscal) subsidy or (direct/indirect) financial aid. It can take the form of an investment premium or a repayable rentless advance. The introduction at the regional level of this increased expansion aid was possible only after a revision of the expansion aid legislation in the early 1990s. The previous legislation (which was of a very general nature and made it possible to grant aid to almost every company on an ad hoc basis, without any sectoral or regional responsibility) was not in accordance with the European legislative framework on state aid (Article 92, EC treaty).

In the Flemish Region, increased expansion aid for environmentally friendly investments by small companies was introduced as early as 1990 (on the basis of the so-called 'ecology criterion'). In the Walloon and Brussels-Capital Regions, similar legislation concerning the use of increased expansion aid for environmentally friendly investments was also adopted in the first half of the 1990s.

The main advantage of increased investment deductions and increased financial aid for environmentally friendly investments is that they can help to convince companies to invest in clean technology and environmentally friendly research and development (OECD 1993). The main drawbacks, however, are that the awarded percentage or the available amount of money is too low (only a 13.5 per cent increased investment deduction and about Bfr3 billion increased expansion aid in the Flemish Region), divided among large companies (the so-called 'Matthew effect'), can be lowered each year and can be spread over (too) many companies. Thus these instruments are usually not decisive factors in the decision about an investment, but only small additional incentives.

Tax differentiation

The use of differentiated tax rates for products and/or services, according to their environmental characteristics, is quite new and fits in the framework of the so-called 'greening of taxation' – shifting the tax burden from labour and capital to environmentally unfriendly products and services (OECD 1993). Interesting fields of action are products and/or services subjected to excise duties and value-added tax. These indirect taxes (and their possible different rates) make it possible to stimulate the use of environmentally friendly products and/or services and to discourage the use of their environmentally unfriendly alternatives. This instrument is an indirect subsidy or financial aid: the consumer avoids paying a (higher) tax and, in effect, makes a profit.

Tax differentiation has as its main advantage that it can be a very effective and efficient environmental policy instrument. Experience has shown that if the difference in tax rate is large enough the results can be tremendous. In Belgium, tax differentiation for environmental goals can be found in the field of excise duties on petrol and heavy fuel, as introduced in the early 1990s (Pittevils 1991). The excise-duty rate on leaded petrol (Bfr21.4 per litre) is (still) higher than the one on unleaded petrol (Bfr19.0 per litre), and the excise-duty rate on heavy fuel with a high sulphur content (Bfr0.25 per kg) is (still) higher than the one on heavy fuel with a low sulphur content (Bfr0.75 per kg).

These differentiated rates have proved to be very effective. For example, there has been a considerable shift from leaded petrol to unleaded petrol in Belgium (see also Chapter 11). However, now that almost every Belgian uses unleaded petrol the federal government is once again reducing the difference in excise duties between unleaded and leaded petrol, but this time to the disadvantage of unleaded petrol, in order to raise revenues.

Another Belgian fiscal initiative concerning cars dates back to the early 1990s: the fiscal stimulation of the anticipated use of clean cars (as suggested by EC Directives 70/220 and 88/76) (K. Deketelaere 1991). The anticipated use of cars with catalytic converters was rewarded by repayment of the traffic levy. This fiscal measure proved very successful.

However, the use of tax differentiation as an environmental policy instrument is still very limited because the scope of possible environmentally unfriendly products, activities and services for which differentiated tax rates (value-added tax and excise duties) can be created is very small. However, change is possible. An ecological revision of the (partially) European harmonised value-added tax – and excise duty – legislation could create many more possibilities for this promising environmental policy instrument.

Environmental labels

The creation in 1994 of a Belgian Committee for the Award of Environmental Labels is the consequence of a European obligation (Regulation 880/92 on ecolabels) (Jadot and De Sadeleer 1992: ch. 9).

However, as early as 1991, the misuse in advertising of unofficial environmental labels (for so-called environmentally friendly products) led to the adoption of some interesting provisions concerning commercial practices and the information and protection of the consumer. On the basis of these, a claim can be introduced with the president of the Tribunal of First Instance, asking for the suspension of misleading (environmental) publicity. A Commission for Environmental Labelling and Environmental Publicity was also created in 1995 (Van Calster 1995).

Taking into account the recent character of most of these initiatives (1991, 1994 and 1995), no serious data are yet available concerning the use, the success and/or the failure of these measures concerning environmental labelling and/or environmental publicity.

Environmental policy agreements

Since the second half of the 1980s a number of voluntary (federal) environmental policy agreements have been concluded in Belgium (Bocken and Traest 1991). Under a (voluntary) environmental agreement between the federal government and industry, the latter commits itself to making special efforts to reduce one or other form of pollution, while the government promises not to adopt new legislation in this particular field (see Chapters 1 and 8). The use of this environmental policy instrument was a consequence, on the one hand, of the growing environmental awareness, commitment and engagement of industry and, on the other hand, the aspiration of the government for voluntary cooperation with entrepreneurs instead of the use of instruments of direct regulation.

These federal (voluntary) environmental agreements all aimed at a reduction of different forms of pollution: for example the reduction (and eventual elimination) of the use of chlorofluorocarbons (CFCs), reduction of the emission of sulphur dioxide (SO_2) and reduction of the production of packaging waste. Although they were quite successful (because of the fact that the goals of these agreements were not that ambitious), these agreements had a kind of 'grey area' existence – their elaboration, adoption and application were not always clear; the impression existed that in most cases the government was a more demanding party than industry and was therefore more easily satisfied with unambitious goals; the impression also existed that industry was often better informed than the government about the (detrimental) environmental impact of products and/or activities, and used that (superior) position in the elaboration and setting of the environmental goals which had to be achieved; a general legal framework concerning these kinds of agreement was lacking.

Because of these problems, which highlighted the public nature of the agreement and the need for ultimate governmental control, the Interuniversity Commission suggested in the early 1990s that a legal framework for environmental policy agreements should be created (IUCHM 1995). This framework was eventually adopted by decree of the Flemish parliament in June 1994, putting an end to the 'grey area' existence of those agreements and the often superior position of industry.

According to this framework (Lietaer 1994; Van Hoorick and Lambert 1995), a voluntary agreement is possible between the Flemish Region, represented by the Flemish government, and one or more representative umbrella organisations of companies, with the goal to limit or prevent environmental pollution, or to promote more effective environmental management. A voluntary agreement cannot replace or be less stringent than the prevailing legislation or regulations. During its validity period, which cannot exceed five years, the Flemish Region cannot issue regulations which are more stringent than those in the agreement. However, the Flemish Region remains competent to issue regulations, either in case of urgent necessity, or in order to fulfil compelling obligations of an international or European legal nature. Similar legislation concerning the use of environmental policy agreements does not exist (yet) in the Walloon and Brussels-Capital Region.

However, since the adoption of this framework only a few (drafts of) environmental policy agreements have been concluded in Flanders. The reason for this is that the legal framework for the adoption of environmental policy agreements is too strict: a very detailed procedure must be followed, both new and former members of the umbrella organisations are bound by the agreement, the decree is of public order, which means that all established legal rules must be followed in detail, and, if they are not, the agreement is null and void. Unless these characteristics of the present legal framework are modified, environmental policy agreements have only a bleak future in the Flemish Region.

Environmental care systems

The goal of an environmental care system is the construction of an instrument at plant level which limits the total environmental burden of a company (K. Deketelaere 1996b; see also Chapter 10). On the European level (EMAS-regulation 1863/93), a lot of attention has been paid to environmental care at plant level. As a consequence, the use of environmental care systems as an instrument in Flemish environmental policy was a suggestion of the Interuniversity Commission (IUCHM 1995) and this led to the adoption of a decree on such systems in April 1995 (K. Deketelaere 1995; M. Deketelaere 1996). In contrast to the EMAS-regulation, however, the Flemish environmental care system is only a partial environmental care system and not an integral environmental care system. The Flemish legislature feared that companies which were not convinced of the merits of such a system would limit themselves to a formal and minimalist application of it, in which case the legal obligation would not achieve its aim. Therefore the Flemish environmental care system was limited to six elements (M. Deketelaere 1996): the appointment of an environmental coordinator; the drafting of an environmental audit; the measurement and registration of emissions and immissions; the drafting of a yearly environmental report; the elaboration of a company policy in order to avoid heavy accidents and to reduce their consequences for people and environment; and the obligation to notify and to warn authorities in case of accidental emissions and disturbances.

It must be said (and this is demonstrated by the EU EMAS) that an integral environmental audit contains much more than these six elements. EU EMAS, which is an example of an integral environmental audit system, for example, also requires, among other things, the drafting of an environmental programme and an environmental declaration.

The above-mentioned environmental audit, an important part of the Flemish environmental care system (Gille 1996), is a systematic, documented and objective evaluation of the policy, the organisation and equipment of an establishment or an activity in the field of environment protection. The environmental audit has to be verified by an external environmental validator. The Flemish government has indicated the categories of establishments for which one (once-only) or more (periodical) environmental audits are required and the elements of which the administration must be notified. Similar legislation concerning the use of environmental care systems does not exist in the Walloon or Brussels-Capital Regions.

The compulsory introduction in the Flemish Region of an environmental care system at plant level can have several advantages (K. Deketelaere 1996b):

- it can reduce the regulatory burden on industry by making the individual company responsible for its own environmental actions;
- it furnishes the environmental authorities with a lot of necessary environmental information;
- it makes better cooperation possible between the companies and the competent authorities.

However, since this system of environmental care at plant level only entered into force in recent months, it is not clear yet whether all of these possible advantages will be realised. Probably, a lot will depend on the (environmental) willingness of individual companies.

It must be said that environmental care systems also could have some disadvantages (K. Deketelaere 1996b). In particular, they create a huge amount of paperwork (audit, registrations, report, etc.) and, above all, quite a few problems of personal liability (see Chapters 1 and 10). Certainly, in the Flemish environmental care debate the penal and civil liability of the appointed environmental coordinator (who has to control the compliance of the company with the environmental legislation) and the leading officials of the company (who have to act on the advice of the environmental coordinator) is the central point of discussion (Faure 1996).

While it is not clear yet whether all of these possible disadvantages will materialise in practice, some have already been strongly emphasised by industry. For example, industry is happy neither with the very broad description of the tasks of the environmental coordinator (for liability reasons) nor with the important role which is foreseen for the Committee for Safety, Health and Embellishment of the Working Places (e.g. will this Committee agree with the person who is appointed as environmental coordinator?).

Deposit, refund, return premium or other collection systems

Deposit, refund, return premium or other collection systems oblige the buyer of a product to pay to the seller an amount of money on top of the price of that product. This amount is paid back to the buyer when he or she returns the product to the seller or an appointed third person. The introduction in Belgium of deposit, refund, return premium or other collection systems as environmental policy instruments is totally connected with the levy on products damaging the environment (K. Deketelaere 1995, 1996a). As indicated above, these systems are a way for industry to avoid the Belgian ecotax by taking responsibility for environmental harm: there are no levies on most of the ecotaxed products (beverage packaging, disposable cameras, batteries, packaging for certain industrial products) when a deposit, refund, return premium or other collection system that meets established conditions is set up. As a consequence, the real goal of the present ecotax legislation is that not one branch of industry pays the levy. If every branch of industry meets its (environmental) responsibility by setting up one or other sufficient collection system (as has already been done, for example, by the battery and disposable camera industries), no branch of industry will pay ecotaxes.

The setting up of deposit, refund, return premium or other collection systems has the advantage that all the producers and consumers concerned can shoulder their (environmental) responsibility. In this way, these systems are a good application of the polluter-pays principle – those who bring polluting products on the market commit themselves to taking those products back after use. This also improves the control and organisation of the management of specific waste streams and makes a final solution more likely.

The application of these systems is still quite limited in Belgium, but they can have a great future when the legislator broadens the field of application for ecotaxes. Although recent in their existence, deposit, refund, return premium and other collection systems are very promising as environmental policy instruments. It is too early yet to speak of significant environmental improvement (e.g. reduction of waste streams) because of these systems, but no manifest disadvantages have emerged for the moment.

In the framework of these systems, one must also consider the federal and regional initiatives which were taken in late 1997 for implementing EC Directive 94/62 concerning packaging and packaging waste (K. Deketelaere 1995). These initiatives concern the establishment of a take-back obligation in order to achieve the use and recycling targets laid down in the agreement between the three Belgian regions on the prevention and management of household packaging waste. Those responsible for packaging are obligated to take back packaging waste that they bring into consumption in Belgium, enough to reach the specified use and recycling targets. The packaging producer can fulfil this take-back obligation itself or invoke a third party (e.g. a recognised organisation).

Furthermore, Article 10 of the decree of the Flemish parliament of 2 July 1981 concerning the prevention and management of waste allows for additional acceptance obligations (K. Deketelaere 1995). For indicated categories of waste, including packaging, the retail dealer, the wholesale dealer and the producer or importer have to accept the waste and the packaging of the products which they have sold, on a 'one-for-one' basis. They can fulfil this acceptance obligation themselves or invoke a third party (e.g. a recognised organisation).

Conclusion

An analysis of environmental policy at the federal and regional levels in Belgium indicates that there is still a dominant use of traditional instruments of direct regulation. However, in recent years these have been increasingly supplemented with other categories of environmental policy instruments, especially instruments of social regulation (environmental policy agreements, environmental care systems) and instruments of market regulation (environmental levies; ecological tax differentiation; environmental subsidies; deposit, refund, return premium or other collection systems).

This Belgian evolution concerning the choice of environmental policy instruments coincides with the European approach as formulated in the Fifth Environmental Action Programme. Instruments of direct regulation will remain the cornerstone of environmental policy, but must be augmented with other environmental policy instruments where this is necessary and/or possible.

References

Bocken, H. and Traest, I. (1991) *Milieubeleidsovereenkomsten* (Brussels: E. Story-Scientia).

De Batselier, N. (1996) *Kiezen tussen Eco en Ego* (Leuven: Uitgeverij Van Halewyck).

Deketelaere, K. (1991) *Milieu en Fiscaliteit* (Bruges: Die Keure).

—— (ed.) (1992) *Recente Ontwikkelingen inzake Milieuheffingen*, LeuVeM, Milieurechtstandpunten No. 1 (Bruges: Die Keure).

—— (1994a) 'De Milieutaks, een Nieuw Instrument voor een Duurzame Ontwikkeling?', in *Fiscalité de l'environnement* (Brussels: Bruylant).

—— (1994b) 'Milieutaks – Start van de Ecologisering van de Belgische Fiscaliteit', *Algemeen Fiscaal Tijdschrift* 45(1): 11–30.

—— (1994c) 'Milieuwetgeving Per Kilo', in S. Gutwirth (ed.) *Milieu Rechtgezet – Een Bezinning over de Grondslagen en Toepassing van het Milieurecht*, Tegenspraak Cahier 15 (Ghent: Mys & Breesch Uitgevers).

—— (1995) 'Het Gebruik van Instrumenten van Marktconforme Regulering in het Afvalstoffenbeleid', in K. Deketelaere (ed.) *Recente Ontwikkelingen in het Afvalstoffenrecht*, LeuVeM, Milieurechtstandpunten No. 6, (Bruges: Die Keure).

—— (1996a) 'Milieutaksen in België – Een Nieuwe Start na een Mislukte Repititie of het Begin van het Einde? – Deel 1', *Algemeen Fiscaal Tijdschrift* 47(9): 288–310.

—— (ed.) (1996b) *Algemene Bepalingen Vlaams Milieubeleid en Bedrijfsinterne Milieuzorg*, LeuVeM, Milieurechtstandpunten No. 9 (Bruges: Die Keure).

Deketelaere, K. and Martens, B. (1994) *Regionale Milieufiscaliteit* (Deurne: Kluwer Rechtswetenschappen).

—— (1996) 'Milieusubsidies', in G. Peeters (ed.) *Subsidiezakboekje* (Deurne: Kluwer Rechtswetenschappen).

Deketelaere, K. and Pittevils, I. (1995) *Federale Milieufiscaliteit* (Deurne: Kluwer Rechtswetenschappen).

Deketelaere, M. (1996) 'Krachtlijnen van het Decreet Bedrijfsinterne Milieuzorg en zijn Uitvoeringsbepalingen', *Milieurecht-Info* 8/9: 1–15.

Faure, M. (1996) 'Strafrechtelijke Aansprakelijkheid, Toezicht en Sancties in het Decreet Bedrijfsinterne Milieuzorg', in K. Deketelaere (ed.) *Algemene Bepalingen*

Vlaams Milieubeleid en Bedrijfsinterne Milieuzorg, LeuVeM, Milieurechtstandpunten No. 9 (Bruges: Die Keure).

Gaines, S. E. and Westin, R. A. (1991) *Taxation for Environmental Protection: A Multinational Legal Study* (New York: Quorum Books).

Gille, B. (1996) 'Milieu-audit in het Decreet Bedrijfsinterne Milieuzorg', in K. Deketelaere (ed.) *Algemene Bepalingen Vlaams Milieubeleid en Bedrijfsinterne Milieuzorg*, LeuVeM, Milieurechtstandpunten No. 9, (Bruges: Die Keure).

Gille, B. and Lambrecht, L. (1996) 'VLAREM II', *Milieurecht-Info* 1, supplement.

IUCHM (1995) *Voorontwerp Decreet Milieubeleid* (Bruges: Die Keure).

Jadot, B. and De Sadeleer, N. (eds.) (1992) *Le Label écologique et le droit* (Brussels: E. Story-Scientia).

Lefebure, B. (1996) 'Kwaliteitsdoelstellingen in het Vlaams Milieubeleid', in K. Deketelaere (ed.) *Algemene Bepalingen Vlaams Milieubeleid en Bedrijfsinterne Milieuzorg*, LeuVeM, Milieurechtstandpunten No. 9 (Bruges: Die Keure).

Lietaer, I. (1994) 'Het Vlaams Decreet over Milieubeleidsovereenkomsten', *Milieurecht-Info* 10: 6–9.

OECD (1993) *Taxation and the Environment: Complementary Studies* (Paris: OECD).

Pauwels, I. and Decoster, P. (1995) 'BBL Heeft Genoeg van Voortdurend Uitstel Ekotaks', *Milieurama* 15(1–2): 16–18.

Pittevils, I. (1991) 'Een Rol voor de Nationale Fiscaliteit in het Milieubeleid', *Documentatieblad*: 197–278.

Steenwegen, C. (1993a) 'Het Ekotaks-akkoord, een Eerste Stap in een Produktbeleid', *Leefmilieu* 16(2): 64–9.

—— (1993b) 'Het Rare Verhaal van de Milieuheffingen en de Ecotaksen', in R. Willems (ed.) *Milieujaarboek 1993*.

Van Calster, G. (1995) 'Het K.B. houdende Oprichting van de Commissie voor Milieu-etikettering en Milieureclame', *Milieurecht-Info* 4: 11–17.

Van Hoorick, G. and Lambert, C. (1995) 'Het Decreet Betreffende de Milieubeleidsovereenkomsten', *Tijdschrift voor Milieurecht* 4(1): 2–10.

Van Humbeeck, P. (1992) 'Milieuheffingen in Vlaanderen: Milieu-economische Reflecties bij de Bestaande Heffingen', *Energie en Milieu*: 201–7.

Van Orshoven, P. (1993) 'De Fiscale Aspecten van de Vierde Staatshervorming', in A. Alen and L. P. Suetens (eds) *Het Federale België na de Vierde Staatshervorming* (Bruges: Die Keure).

VEV (1995a) *Groeiprioriteiten voor Vlaanderen* (Antwerpen: Vlaams Economisch Verbond).

—— (1995b) *Cahier Milieu* (Antwerpen: Vlaams Economisch Verbond).

Steering and Working Group on Environmental Levies (1993) *Milieuheffingen: Een Instrument voor een Vlaams Milieubeleid* (Brussels: VUB-RUCA; Free University of Brussels, Faculty of Economics).

6

NEW ENVIRONMENTAL POLICY INSTRUMENTS IN SPAIN

Susana Aguilar Fernández

Introduction

Spain has traditionally exhibited a weak – and in certain areas non-existent – environmental policy. This situation, together with a predominance of factors (discussed below), has hitherto limited the introduction of new environmental policy instruments.

Where it existed, Spanish environmental policy was characterised by the low relevance of NGOs and industrial groups, and by the absence of any significant effort at harmoniously linking economic policy and environmental protection (Aguilar 1993). Despite this, four major sources of pressure can be identified, each of which encourages the emergence of a debate about (and the putting into practice of) new environmental instruments. These pressures can be broadly categorised as: accession to the European Union (EU), which imposed enormous environmental costs and prompted a search for less expensive tools; growing environmental awareness in Spain, which led to calls for more environmentally effective instruments; changes in the traditional exclusive policy style, promoting greater involvement of social groups in the domestic policy process; and administrative reforms, which encouraged greater coordination of environmental policy.

Before its accession to the EU Spain had naturally had different pieces of environmental legislation (the first and most important one being the Law of Air Protection, which dates back to 1972), but their enforcement was not prioritised because the policy was neither socially nor politically relevant. Entry into the EU helped to upgrade environmental issues, not least because of the obligation to implement a high number of Community Directives. This facilitated changes in both the public apparatus, so that environmental policy could be better coordinated, and the national policy style, so that the promotion of some cooperation amongst the actors involved in the policy could ease its implementation. All of this was accompanied by the gradual emergence of environmental concerns, which furthered the establishment of a (still embryonic) debate about new instruments in the policy, be they voluntary agreements or economic and fiscal tools.

Spain's accession to the EU

Shortly after accession to the EU, Spain was very busy attempting to catch up with the bulk of environmental policy which was suddenly enforceable. This effort can be explained not only by the (high) quantity and (stringent) quality of environmental Directives and the relatively underdeveloped stage of Spanish legislation, but also by the fact that the government did not strike any important deal on environmental issues in the negotiations leading to Community membership. Since 1986, therefore, the country had to transpose into its domestic law (and, more importantly, to implement) a large number of stringent Directives without any concessions concerning transitional periods or special conditions of application.[1] Until that moment, the law had tended to ignore the harmful effects exerted on the environment by public or private activities and projects, while the authorities had rarely shown any special interest in the matter – environmental crimes, for instance, were not included in the Spanish penal code until 1993, and, even then, these were simply limited to some punitive actions concerning discharges and emissions. Only at a later stage did the country begin to offer some resistance to certain aspects of EU environmental policy which allegedly entailed disproportionate costs or responded to the political priorities of other (mainly northern) member states.[2]

On the whole, Spanish environmental regulation has mainly revolved around water policy, undoubtedly due to the long-standing legislative tradition in the management of a crucial and scarce asset. Not only has there been little legislation in relation to waste, coasts and chemical products, but also the few existing pieces of regulation have basically followed a command and control approach. For this reason, the Law of Water, passed one year before Spanish EU membership, in 1985, represented a watershed because it introduced a tax on discharges called 'canon'. This tax was meant to help finance the costs associated with the cleaning-up process of rivers (and the construction/upgrading of the public sewage system in the case of indirect discharges) or to promote the installation of purifying plants at the factory level (for direct discharges with no sewage system in place). Guidelines as to the type of plants and the specific technology needed were basically missing, although public financial aid (in the shape mainly of soft credits) was offered for the modification of infrastructures and techniques that would lead to a reduction of both water consumption and water pollution.

Achieving the high standards of legal protection of the environment contained in the Community's regulations from such a base was therefore a very difficult and expensive task. In 1995 an exhaustive report made by the

Directorate-General of Environmental Policy quantified the necessary investments for the correction of the Spanish environmental deficit at 5.3 billion Pta over the next ten years (DGPA 1995). This figure comprised the deficit accumulated due to delays in applying EU environmental legislation and the costs derived from the application of new Directives (such as 91/271 on the treatment of urban waste water). This estimate meant that the effort made hitherto would have to be more than doubled, starting from the fact that the current environmental expenditure comprised only 0.73 per cent of GNP.

Growing realisation of the enormous expenditure required to meet EU environmental standards, implementation deadlines which had long since expired and continuous pressure from European Court of Justice (ECJ) rulings led to the conviction that profound modifications were urgently needed. This drew attention to the potential advantages offered by new and less expensive (market-based) policy instruments. Among others, this novel focus began to be promoted by the former Ministry of Public Works, Transport and Environment (MOPTMA), whose minister, José Borrell (an engineer with a degree in economics), had on several occasions adhered publicly to the need to introduce more efficient economic instruments in environmental policy. In 1993, for instance, the official bulletin of the MOPTMA, *Información de medio ambiente*, included an editorial in which environmental economic instruments were discussed at length. In the same vein, the head of the recently created Ministry of the Environment (MIMAM), Isabel Tocino, expressed her willingness to foster those instruments and to allow a larger leeway for private entrepreneurs in her first public appearance at the Chamber of Deputies in May 1996 (*Journal of Sessions of the Chamber of Deputies* 1996). More precisely, the minister prioritised the gradual application of the polluter-pays principle,[3] the analysis of ecological taxes, the passing of as many financial and economic incentives as necessary for a policy aiming to accomplish the principle of sustainable development, and the promotion of environmentally friendly investment by firms so that those who reduce pollution will be economically rewarded. Concerning the latter measure, the idea of signing agreements with financial institutions for the concession of credits to entrepreneurs, as well as the introduction of 'special negative contributions' which would allow compensation for those expenses voluntarily incurred by individuals to protect the environment, was conceived (*Información de medio ambiente* 1996, no. 43). These political statements have also been flanked by the organisation of a growing number of public debates about economic instruments. Examples of these can be found at the annual International Fair of the Environment (PROMA) in Bilbao, which brings together private entrepreneurs from all over the world and public authorities, and at the Environmental Forums promoted by the former DGPA. Likewise, the Office for the European Communities in Spain has also organised workshops where the new trends in EU environmental policy have been debated, as it did in 1992 when the topic of financial tools for ecological improvement was chosen for discussion (*Boletín informativo de medio ambiente* 1992, no. 9).

To some extent, the inter-party consensus about the need to reorient environmental policy implicitly recognises the compliance problems associated with the command and control approach, and the weak influence of (scarce and poor) fines on environmentally harmful practices. Moreover, the traditional regulatory approach entails high administrative costs, which are difficult to allocate in a federal-like state such as Spain.

Increasing environmental awareness

Since the mid-1980s opinion polls have reflected that environmental protection was gradually becoming a central objective at the societal level, slowly replacing what had previously been a weak interest in the subject. However, the alleged concern about the environment has not clearly translated into social willingness to act in favour of it or into greater support for green political parties. Politically, the new centrality of the environment can be accounted for mainly by EU membership and the concomitant need to transpose and implement a bulk of stringent legislation.

The salience of this topic was exemplified by the speech given by José Borrell at the Chamber of Deputies in 1993, in which he spelled out for the first time the priorities (soil erosion, optimisation of water usage, waste and urban-environment quality) and methodologies of Spanish environmental policy (*Journal of Sessions of the Chamber of Deputies* 1993). However, in spite of this advance, Spain continued to be 'different' for some time because of the lack of a ministry for the environment.

For years, the creation of this ministry had been demanded by various social sectors and political parties but the governing Socialist Party (PSOE), which was in power between 1982 and 1996, always opposed it on the grounds that the 1978 Constitution had already distributed environmental powers amongst the different political-administrative levels, making the setting up of the department redundant. However valid this argument may be, the absence of this agency left Spain as the only EU member state which did not have any ministry of this type at a time when the European Commission was repeatedly blaming the environmental deficit on intergovernmental coordination problems. This absence was also referred to by industry, which found in it a good excuse for its too frequent negligent behaviour in environmental protection.

The electoral triumph of the Popular Party (PP) in March 1996 brought with it the setting up of the first Spanish Ministry of the Environment. While in most cases its political priorities do not differ dramatically from those expressed by the previous government, its creation has had at least two positive effects: the establishment of a specific parliamentary Commission for the Environment, which will undoubtedly facilitate the formulation and monitoring of this policy, and the concentration of environmental powers in the new agency. The short time-span which has elapsed since this ministry began to function does not, as yet, allow any conclusive evaluation of its performance. For this reason, it is not clear at all whether the new agency has simply paid lip service to the cause of the environment or whether, quite to the contrary, it is trying to modify long-standing institutional inertia and harmful practices. In any case, one thing which stands out is the more frequent reference made by the responsible figures within the ministry to economic and voluntary instruments, so that entrepreneurs are provided with incentives (in terms of tax reductions, flexible deadlines, access to state aid, etc.) to comply with environmental targets. Moreover, the minister of the environment has publicly recognised that environmental law has become too abundant and complicated, and that private actors have a terrible time when trying to decipher and make sense of it. All this is naturally linked to the new 'ideological mood' of the PP, a party which is further promoting market-based instruments, deregulation and privatisation processes in different fields. (For instance, the 1985 Law of Water has been reformed, amongst other things, to give groundwater back to private owners.)

Changing state–society relations

Environmental policy-making in Spain has mainly been conducted by public actors (government and civil servants), whereas private ones (industrial groups, and, above all, environmentalists and citizens) have played a secondary role in the process. Not even independent experts, in spite of the highly technical nature of this policy, have been regularly called upon; this participation would have run counter to the usual practices of an administration that rarely sets up advisory bodies or experts' committees, preferring, on some occasions, to contract them individually or on a case-by-case basis (Martín Rebollo 1984). As a consequence of this, no institutionalised relationship or formal cooperation has been established between public authorities and social groups for many years. This situation has overloaded the state with numerous tasks and has aggravated implementation deficits because the political and institutional arrangements which promote cooperation between public and private actors have better policy records than those of a hierarchical and non-inclusive nature (Aquilar 1997a).

In spite of the non-cooperative policy style which has traditionally prevailed in environmental policy-making, several projects endorsing the creation of organisations which would allow for social participation have been discussed over the last decade. In most cases, however, these projects were not approved, while the rare organisations which were finally set up rendered very limited results. In 1984 one of the many General Law of the Environment bills under discussion envisaged the creation of a National

Commission composed of social groups and public institutions, but this bill has been abandoned since then (Costa 1985); the project on Basic Guidelines for the Protection of the Environment, which recommended the setting up of a High Council of the Environment in which administrative staff, Members of Parliament (MPs), environmentalists, citizens, economic and professional associations would be brought together, was not finally taken up either (*Boletín informativo de medio ambiente* 1980, no. 14). The only organisation of this type which was eventually put into practice was the Committee of Public Participation (CPP), established in 1983 and suppressed three years later. The CPP, which intended to get environmental groups involved in the administration, did not manage to play an outstanding political role because its decisions were not binding on the government (*Información ambiental* 1986, no. 9). The latest case of institution-building regarding social participation in the environmental policy process is the Advisory Council of the Environment (ACE), set up in 1994.[4]

The ACE, an organisation upon which two basic functions have been conferred – to give advice on certain laws, bills and programmes, and to make reports and proposals at the request of the administration or on its own initiative – is comprised of thirty-six people who belong to a wide array of sectors (administration, environmentalists, farmers, business people, consumers, citizens). This organisation has exhibited an ambivalent functioning to date: on the one hand, it has turned out to be a good forum of debate; on the other hand, and according to some of its participants, it has enjoyed only weak leverage on the government stance.[5]

In the new regional administration the situation concerning social participation in environmental policy is similar to the one described for the state level. Most units entrusted with environmental protection (irrespective of their administrative status and name, agencies, departments, councils) do not count on interest groups but merely on representatives of the administration (Suárez 1990). Those rare cases in which individuals not belonging to the public apparatus have been allowed to participate have been created recently (making their assessment rather premature) or have functioned disappointingly: the Council of the Environment in Navarre, for instance, was set up in February 1993 but disappeared three months later, whereas the Advisory Council of the Environment in Andalusia has only very rarely been convened since its foundation. All this has culminated in the absence of an institutional design in which discussion about new policy instruments in environmental policy, along with other topics, could take place. That is, the lack of appropriate forums where the main participants in this policy could gather together has traditionally prevented the emergence of an important debate on economic tools in the field of environmental protection.

Yet public authorities (be they state or regional) are becoming increasingly convinced of the need to incorporate private actors into the making of environmental policy. In this sense, the former Secretary of State for the Environment, Cristina Narbona, was well rated by environmentalists because of her cooperation-prone attitude, as well as her willingness to get conservationist interests involved in different forums and institutions concerned with the protection of the environment. As regards the new government, the MIMAM has declared her intention of widening the social basis of the ACE (now renamed Advisory Forum of the Environment) so that people involved in, among other things, land planning, hunting and fishing activities can participate in it. Likewise, the new ministry has proclaimed that it will promote the participation of social and economic actors for the improvement of the environment, and will foster dialogue with social groups and NGOs from the perspective of 'shared responsibility'. Whether this new process of institution-building will lead in the near future to a significant degree of cooperation between the administration and the social parties interested in environmental issues, gradually transforming the traditional non-cooperative policy style, or whether it will simply pay lip service to recommendations put forward by different institutions (such as the European Commission) remains to be seen. However, one thing is definitely established by now: the reluctance of public authorities regarding social participation is diminishing and new policy practices which permit private actors to play a more important political role are being put into practice. One of the most outstanding examples of these new practices is voluntary agreements (discussed below).

Administrative reforms

The state environmental administration has traditionally been characterised by its dispersion of responsibilities among different ministries and between different levels of political authority (Aguilar 1997b). In an attempt to resolve this situation the Interministerial Commission for the Environment (CIMA) was set up in 1972. A number of things (overload of tasks, non-compulsory reports, lack of funds) impaired the functioning of this organisation until its disappearance in 1987. Alongside the languishing development of CIMA, the Ministry of Public Works (MOPU; converted into MOPT, Ministry of Public Works and Transport, in 1991, and, two years later, into MOPTMA) became the main agency for the protection of the environment in 1977 and started, consequently, to concentrate an increasing number of tasks in this policy area.

The entry of Spain into the EU has fostered the putting into practice of several projects that aimed at overcoming both the traditional problems of dispersion of environmental responsibilities at the central level and the new problems of vertical coordination between state and regional administrations. In 1994 a new project which would create an interministerial commission to coordinate the environmental competencies among the different sections within MOPTMA, and between this department and the Ministries of Agriculture, Industry,[6] Education and Justice was announced (*Información de medio ambiente* 1994, no. 24). A year later, the top state and regional officials in this policy area were convened at a Sectoral Conference of the Environment, and in 1991 this forum adopted common and immediate goals for the protection of the environment (*MOPU informa* 1991).[7]

The MOPTMA period between 1993 and 1995 represents an attempt to configure more consolidated environmental policies. For the first time, the ministry proposed the establishment of a discussion (within and outside of the Chamber of Deputies) about horizontal models for environmental integration, particularly in economic terms. Coordination between environmental policy and economic policy was constantly sought throughout this period, and novel debates concerning the use of market mechanisms, the imposition of prices which would internalise the real environmental costs of products, and the possibility of green tax reform took place at the parliamentary level. More specifically, in 1993 the leading official for MOPTMA, José Borrell, presented at the Chamber of Deputies a policy design which would eventually lead to the elaboration of a National Plan for the Environment (MOPTMA 1993). Within this design, a heading entitled 'the reorientation of market mechanisms' was included. It embraced, among other things, the following: the establishment of a pricing system to internalise recycling or waste-treatment expenses into the price of goods, support for agreements between the administration and entrepreneurial sectors to introduce environmentally friendly technologies, the need for the public sector to set examples in environmental policy, the introduction of changes in the fiscal system so that different activities would be taxed depending on their polluting effects or intensive use of non-renewable energies, and the promotion of ecoauditing at the firm level. The need to apply pricing mechanisms was exemplified by Borrell by means of a particular public good: water. He highlighted the fact that citizens were not paying the total costs associated with water treatment for domestic uses and that the real costs of the process should be somehow reflected in the bill.

In spite of the four changes discussed above, the issue of economic instruments in environmental policy has not yet been extensively dealt with. Environmentalists have basically focused on denouncing illegal practices and enforcing the current legislation, whereas industry, although it has paid lip service to a market-oriented strategy, has not shown a big interest in new tools because the traditional command and control approach is not rigorously applied anywhere, so that entrepreneurs have hitherto incurred only limited costs. Political parties, for their part, have not devoted much time to the discussion of this topic because it does not enjoy social salience. For instance, when the PP was still in opposition it defended economic instruments as a sort of theoretical/ideological device without knowing whether they would be feasible or beneficial at all. Even now, the numerous public statements in favour of them have not translated into a serious discussion about their utility, or into specific legislative measures or public plans.

New policy contents: economic and voluntary instruments

Spanish environmental policy has generally had a regulatory character because it has imposed, through legislation and regulations, limits and prohibitions on the productive activities of business, as well as on the consumption practices of citizens and social groups. Increasingly, however, in response to the various pressures discussed above, this command and control approach has begun to work alongside other types of instruments: self-regulating and voluntary instruments, which allow individuals and groups to anticipate the regulation and to adopt, voluntarily, certain policy objectives before the authorities approve and enforce them; and economic and fiscal instruments, which pursue the accomplishment of those objectives by means of market mechanisms and cost-efficiency measures. The most outstanding economic and fiscal instruments are discussed below (Aquilar 1998).

Environmental insurance

In general terms, environmental insurance or environmental liability refers to the obligation placed on those responsible for certain activities with harmful effects on the environment to take out an insurance policy to cover potential environmental risks and damage. This scheme has traditionally been impaired because the increasing demands for insurance coverage have not been matched by a sufficient supply of resources from insurance companies.

In the 1990s a group of experts founded a Spanish Pool of Pollution as a type of environmental insurance. This pool seemed quite ambitious at first glance since it intended to cover the incidental and gradual effects of pollution and a wide range of other issues such as: environmental restoration, prevention costs, substance-elimination costs, extraordinary emergency costs, and judicial and bail costs.

The aim of this instrument at that time was to meet the requirements of coverage which arose as a result of the application of the Law of Toxic and Hazardous Waste (Law 20/86), which embraced a highly demanding compulsory insurance policy. The new MIMAM announced at the end of 1996 that an environmental liability bill was being discussed and was likely to be approved soon. This bill envisages, among other things, that a new insurance policy will be compulsory on all the activities which may harmfully impinge upon the environment, not only on those related to toxic waste (as was previously the case), and that this insurance must be signed before the affected firms can be authorised (*Información de medio ambiente* 1997, no. 47).

Ecotaxes

On the state level, the first experience with ecotaxes was related to water policy. The aim of the 1985 Water Law was to introduce fiscal measures for all water uses, but hitherto only domestic discharges and industrial uses under 3,500 m^3 per year and firm-based discharges have been subject to taxes. The amount of tax is established on the basis of two elements, one fixed and linked to resource use, the other variable and linked to water consumption and the type of pollutants discharged. The money raised is basically meant for corrective clean-up measures and the improvement of the public sewage system. In this sense, the new instrument has basically functioned as a revenue-raising tool which has complemented the predominant command and control approach, and it has not seriously promoted the reduction of pollution by means of economic incentives or the specification of technological requirements.

Despite the implementation of a water tax, a workshop on the Global Management of Water held at the Chamber of Deputies in 1990 depicted a worrying situation in this policy and recognised the need to promote clean-up programmes as well as applying the water tax to finance them. It was estimated that 40 per cent of firms have not been charged the tax (approximately 6.5 billion Pta) and 70 per cent of the municipalities have never paid it (18.8 billion Pta). Four years later the headway made in water policy was insignificant and MOPTMA was still trying to implement the tax in those regions which did not have this instrument (that is, Andalusia, Canarias, Cantabria, Castille-La Mancha, Extremadura, Basque Country, Aragón, Asturias, Castille-León and Murcia). The only region which has been charging this tax and increasingly receiving more money is Catalonia: 98 billion Pta in 1983, 287 billion Pta in 1984, 663 billion Pta in 1985, 844 billion Pta in 1986, 1 trillion Pta in 1987 and 1.2 trillion Pta in 1988 (JSC 1990). Also in 1995, MOPTMA estimated that just 5 per cent of the existent 56,400 direct discharges – out of a total of 300,000 points of water discharges – were authorised. Moreover, and to make things even worse, public aid to promote the introduction of the prevention principle in this policy has been very limited up to now.

To further water pollution control a programme called SAICA (automatic system for the monitoring of water quality) was approved in 1993, with an estimated investment of 10 billion Pta between 1993 and 1996 – 85 per cent of the budget coming from the EU cohesion fund. At the beginning of 1995 the government passed the National Plan for the Cleaning of Rivers, which, with a budget of 1.7 billion Pta, will strive to clean up the water of 60 per cent of the population.

Likewise, the fines for breaches of the Water Law have recently been updated and now reach a maximum of 75 million Pta. Over the last few years about 12,000 notices, entailing fines which amount to 5 billion Pta, have been levied on firms. The new MIMAM, for its part, has announced that it will emphasise the need to enforce the current water financing system and that it 'will dictate a decree incorporating the doctrine established by the Central Economic and Administrative Court [so that] those who have to pay [the water tax], will receive their settlements technically and juridically justified' (*Información de medio ambiente* 1996, no. 43).

Unlike those for water, the establishment of charges on waste oils represents the only consolidated experience of ecotaxes in Spain. The origin of this tax is found in a number of corresponding EC Directives (e.g. 75/439 and 87/101) and the national regulations dealing with toxic waste of 1989 and 1990. This legal framework envisaged the possibility of granting indemnities to the enterprises which carry out the collection and/or disposal of waste oils, so as to support their correct management and treatment. This compensation would be financed through a tax applied to those products which were more or less transformed into waste oils after their utilisation. Although these regulations were correctly applied and abided by, a bill was prepared in 1995 because 'the expected results in relation to the decrease in discharges had not taken place, perhaps due to the absence of sufficiently clear and stable financial mechanisms for the internalisation of external economies'. In November 1995 a Law on Waste Oil, which established a management scheme for this subproduct complementing that already envisaged by the 1986 Law of Toxic Waste, was eventually passed.

This new law embraced a tax on industrial waste oils which would permit the administration to dispose of the necessary funds to compensate firms for the costs incurred by the fulfilment of Community requirements. This tax, which was indirectly applied to specific types of consumption, charged the manufacturing, acquisition and importation of oils at the rate of 6 pesetas per kilogram. It was estimated that the money raised by this levy would amount to 2.4 billion Pta annually. Payment to the affected firms would be carried out by means of agreements with their respective regional governments. Furthermore, the regions would undertake environmental programmes in this field (compensation, aid to entrepreneurs, information campaigns and investment in research and development (R&D)) on the basis of those revenues (*Información de medio ambiente* 1996, no. 40; the implications of recycling tax revenue are discussed in Chapters 1 and 11).

Concerning the issue of containers and packaging, the first discussions about a bill which contained relatively stringent objectives and innovative instruments produced harsh confrontation between businessmen and the government in the mid-1990s.[8] For instance, MOPTMA and the Spanish Confederation of Entrepreneurs (CEOE) entered into an angry discussion at the end of 1995 about the utilisation of PVC (polyvinyl chloride) plastics.

The ministry announced that it was going to reduce their use – following the example set by other Community countries – and openly criticised industry's resistance to this measure. In addition, the Secretary of State for the Environment, Cristina Narbona, pointed out that 'in Europe many firms had voluntarily done what the CEOE considered impossible' (*El País*, 23 November 1995). Yet the main problem arose when the industry accused the administration of withholding from the parliament and the public a report, financed by MOPTMA, which allegedly showed that concentrations of this product in Spain were below those leading to acute alterations. Replying to this accusation, the Director of Environmental Policy stated that legal action concerning the case could not be ruled out 'because the misuse of that report was very worrying' (*El País*, 23 December 1995). Disputes like this would explain the following statement of the director of the European Agency of the Environment, Jiménez Beltrán, in 1996: although 'coordination and cooperation . . . are vital in environmental policy, in Spain the different participants to [policy discussion] do not accept each other, and there is a mutual distrust' (*El País*, 16 February 1996).

The new administration of the PP finally approved a Law of Packaging Residues in 1996 which has considerably reduced previous targets and has consequently encountered widespread criticism from environmentalists. This law envisages that the different agents participating in the commercial chain of a packaged product must charge their clients, up to the final consumer, a sum of money which will be given back to them once the package is returned. It is also foreseen that if the affected agents want to be exempted from this obligation they should participate in an integrated management system which guarantees the regular collection of these subproducts, as well as of the recycling and valorisation[9] targets applied to them. These systems will be implemented by means of voluntary agreements which will be supervised by the regional governments. The targets embraced, which will have to be fulfilled before the middle of the year 2001, are as follows: valorisation of 50 per cent minimum and 65 per cent maximum (by weight) of all packaging; and, within the previous objective, recycling of 25 per cent minimum and 45 per cent maximum. Finally, the law allows for the regions to establish economic and fiscal measures if the minimum targets are not reached.

On the regional level the introduction of economic and fiscal instruments in environmental policies has not gone much further than on the state level. On the whole, only two regions have elaborated strict ecotax regulations, aimed at levying installations and activities with environmental harmful effects. The first of them is the Balearic Islands, where in 1991, to raise money, a 1 per cent tax was placed on firms involved in the production, storage and transformation of electrical energy and fuels.[10] The other regional government which established a rate for atmospheric pollution was Galicia, in 1995.[11] In this case, the act which envisaged a tax on sulphur-dioxide (So_2)

emissions and oxides of nitrogen has been debated and agreed upon by social agents together with the authorities. The intention of this levy is not, unlike the first example, tax collection but, rather, a direct reduction in atmospheric pollution. Although the application of this system seems to be satisfactory, not enough time has elapsed since its introduction to allow a practical evaluation of its full effects.

Ecoauditing

Environmental audits are an interesting contribution to sound environmental management because they are often a prerequisite for companies which want to apply for public grants and subsidies. These companies must carry out certain measures at plant level if they want to benefit from grants and to get an environmental certificate. In most cases, the application of audits is entrusted to regions, although some cases fall under the responsibility of the state administration (*Información de medio ambiente* 1995, no. 36). Despite their importance, audits are evolving in a slower and less orderly manner than was initially expected.

At a recent National Congress of Environmental Law an expert on the subject remarked that the greatest volume of business in the Spanish auditing sector centred around training courses. There are more conferences and congresses held on the subject than there are companies carrying them out. Moreover, the few audits carried out are riddled with significant confusion: first, it is seldom remembered that audits are meant to analyse the environmental-quality systems used by companies rather than the production processes which are at the heart of business activity (unfortunately, public administrations often and mistakenly refer to a company's implementation of legislation affecting its production processes as 'auditing'); second, certain aspects of ecoaudits prevent them being used by entrepreneurs since they frequently uncover non-compliance with regulations, and this results in the issuing of sanctions by the authorities and even in criminal proceedings (see Chapters 1 and 10).

Voluntary agreements

Alongside these economic instruments, voluntary measures which aim to adjust industrial behaviour to EU environmental obligations by means of negotiating agreements with different economic sectors have also been introduced. These compromises, which never alleviate or help circumvent requirements in force or to be enforced, can include objectives which go beyond EU targets. If this is the case, firms are given special treatment in terms of public aid, soft credits and extended deadlines. However, non-compliance with agreements by industry can prompt government intervention. In this sense, the authorities, which would otherwise simply act as a watchdog, can resort to imposing policy objectives through the usual regulatory approach (enacting decrees, passing laws) (see Chapters 1 and 8).

The first agreement to be signed was in 1989 and concerned MOPU and the Spanish Association of Sprays, with the aim of reducing the use of chlorofluorocarbons (CFCs). Most of the compromises of this type were signed in the mid-1990s and they concern the following: the agreement between the government and battery producers to reduce the mercury content in their products; the agreement between the Government Department of the Environment and some manufacturers on the recycling of paper and cardboard; talks about a management programme for used tyres between the General Subdirectorate of Waste and industries linked to this field; and negotiations between the administration and various packaging sectors (glass, plastics, cans) to promote recycling (*Información de medio ambiente* 1993, no. 17; *El País*, 16 December 1994). There are also cases in which the authorities have supported, without directly participating in, accords between private actors: for instance, the state secretary for environmental policies has backed an agreement between the vehicle manufacturers' association (ANFAC) and the firm REYFRA to carry out a recycling project in the motor-car sector.

Finally, the Ministry of Industry's Programme for the Creation of a Technological and Environmental Basis (1990–4) should also be mentioned. Although this programme cannot be defined as a new environmental policy instrument, it does represent the most important financial effort (38 billion Pta) ever made by the administration to help industry comply with EU requirements. The main objectives of this programme were to overcome the environmental deficit of Spanish industry (estimated at 5 billion Pta) and to support the (embryonic) green domestic market: 35 per cent of the applications have gone to air protection, 61 per cent to water-pollution abatement, and money has also been given to R&D (above all to projects which could show the utility and feasibility of certain techniques and programmes). One of the most remarkable features of this programme has been the adoption of a precautionary approach in a policy area where most measures have hitherto had a curative character. Thus, clean technologies, as well as the implementation of standards which are more stringent than those set up by legislation, have been strongly promoted – in fact, 50 per cent of the firms which have received money from the programme fall within this category. The Programme for the Creation of a Technological and Environmental Basis also embraces a section for the environmental adaptation of twenty-seven large firms whose investment needs amount to more than 1.5 billion Pta.

Conclusion: a tentative evaluation of the new instruments in environmental policy

It is still too soon to decide whether the new tools discussed have been beneficial to the environment. On the one hand, instruments such as the Ministry of Industry's programme have seemingly had positive effects in terms of making industry gradually aware of the need to abide by EU legislation, without resorting to delaying or obstructing strategies to the same extent as before, and in terms of convincing some sections of the business community of the advantages associated with environmentally friendly production. The results of the voluntary agreements have been ambivalent, but they have at least helped to remove some of the reluctance which public authorities traditionally exhibited about social participation. Equally, these agreements have increased the interest of some private groups in being formally involved in the environmental policy process. On the other hand, however, there are instruments which have clearly failed to accomplish their objectives, as is shown by the case of the water tax. Despite the uncertainties which surround the evaluation of new instruments in Spanish environmental policy, one thing can be taken for granted: the sheer fact of experiencing new approaches with the (explicit or implicit) aim of improving environmental protection is in itself positive. Yet the problem is that institutional inertia and a deep-seated distrust between the actors still pervades this policy.

Notes

1 In fact, there was only a concession concerning unleaded petrol.
2 All this explains why Spain, along with other southern EU member states, insistently demanded that more EU resources should be allocated for environmental protection, first, by pressing for an increase in the money assigned to the LIFE programme (financial instrument for the environment) and, second, and more importantly, by leading the campaign in favour of the creation of a new financial instrument connected with the principle of cohesion: the Cohesion Fund.
3 Spain, like most countries, adheres to the polluter-pays principle although, in fact, most resources assigned to environmental protection are public. Despite this, public investment (as a percentage of GNP) is small, above all when compared with the money spent by other OECD countries: in the mid-1980s pollution-control expenditure amounted to 0.89 per cent of GNP in Canada, 0.60 per cent in the United States, 1.17 per cent in Japan, 0.56 per cent in France, 0.78 per cent in Germany, 0.95 per cent in Holland, 0.66 per cent in Sweden and 0.62 per cent in Great Britain, whereas in Spain, in 1987, total public investment in the environment (not only in pollution abatement) was only 0.60 per cent of GNP. Yet this percentage has been growing: from 0.64 per cent in 1988 to 0.72 per cent in 1989 (OECD 1991; MOPT 1991).
4 The National Commission on Climate Change, set up in 1992 to comply with the compromises reached at the International Summit on the Environment and Development organised in Rio, will not be included in the analysis because it is composed exclusively of scientists.

5 This fact would explain why Greenpeace and Aedenat (an environmental association based in Spain) left ACE in the summer of 1995 (*El País*, 20 June 1995).
6 Likewise, an agreement between MOPTMA and the Ministry of Industry to cooperate on environmental issues and to coordinate the Spanish stance in the EU was signed in 1994.
7 The equivalent to this conference in conservation policy would be the National Commission for Nature Protection (NCNP), set up in 1994. NCNP operates at two different levels: one of general directors and another of experts who comprise four different technical committees (natural areas, flora and fauna, wetlands and forest fires).
8 This bill was preceded by the approval of a National Plan for Toxic Residues in 1995 which aimed to minimise 40 per cent of the waste by the year 2000.
9 Valorisation is defined as every procedure (including incineration) which allows advantage to be taken of the resources incorporated in the packaging subproducts.
10 Valuation of the various worths compounding the tax base is done by capitalising 40 per cent of the average exploitation gross revenue obtained from turnover in the previous three fiscal years. This figure is used as the basis from which to accrue 1 per cent, which is the amount payable. See Act 12/91 of the Autonomous Community of the Balearic Islands, relating to the creation of a tax on installations with effects on the environment.
11 See Act 12/95 of the Autonomous Community of Galicia, relating to the creation of a tax on certain atmospheric polluting activities.

References

Aguilar, S. (1993) 'Corporatist and statist designs in environmental policy: the opposing roles of Germany and Spain in the Community scenario', *Environmental Politics* 2(2): 223–47.

—— (1997a) *El Reto del medio Ambient. Canflictos e Intereses en la Política medioambiental Europea* (Madrid: Alianza Universidad).

—— (1997b) 'Subsidiarity, shared responsibility, and environmental policy in Spain', in U. Collier, J. Golub and A. Kreher (eds) *Subsidiarity and Shared Responsibility: New Challenges for EU Environmental Policy* (Baden-Baden: Nomos).

—— (1998) 'Las políticas de medio ambiente, entre la complejidad técnica y la relevancia social', in R, Gomá and J. Subirats (eds) *Políticas Públicas en España* (Barcelona: Ariel).

Boletín Informativo de Medio Ambiente (1980, 1992) (Madrid: MOPU).

Costa, P. (1985) *Hacia la Destrucción Ecológica de España* (Barcelona: Grijalbo).

DGPA (1995) 'El Déficit de inversiones ambientales', General Direction of Environmental Policy (DGPA), MOPTMA. Unpublished document, November.

El Pais (various issues). Daily newspaper, Madrid.

Información ambiental (1986) (Madrid: MOPU).

Información de medio ambiente (1993, 1994, 1995, 1996, 1997) (Madrid: MOPTMA).

JSC (1990) Junta de Sanejament de Catalunya, *Memoria 1988/89* (Barcelona).

Journal of Sessions of the Chamber of Deputies (1993) transcription of the session of the Commission of Infrastructures of 24 November 93, relating to the appearance of the Minister of Public Works, Transport and Environment.

—— (1996) transcription of the session of the Environmental Commission of 24 June 96, relating to the appearance of the Minister of Environment.

Martín Rebollo, L. (1984) 'Las relaciones entre las administraciones públicas y los administrados', in J. Linz *et al.*(eds) *España: un presente para el futuro* (Madrid: Instituto de Estudios Económicos).

MOPT (1991) *Medio ambiente en España 1990* (Madrid: MOPT).

MOPTMA (1993) 'Líneas basicas del diseño de la politica ambiental' [Basic lines in the design of environmental policy], State Secretary for Housing and the Environment (Madrid: MOPTMA).

MOPU informa (1991) (Madrid: MOPU).

OECD (1991) *The State of the Environment* (Paris: OECD).

Suárez, A. (1990) 'Visión general de la problemática asturiana sobre el medio ambiente', in Ernst and Young, *La Estrategia de las Empresas e Instituciones Asturianas para el Cumplimiento de la Normativa sobre el Medio Ambiente* (Oviedo: Ernst and Young).

7

NEW ENVIRONMENTAL POLICY INSTRUMENTS IN ITALY

Alberto Majocchi

Introduction

Italian environmental policy has traditionally been marked by a commitment to direct regulation, through a system of command and control measures. The impetus for regulation in certain sectors was strictly domestic, reflecting increasing concern for environmental protection. In other sectors, where domestic support was lacking or legislation insufficient, increasingly stringent Italian environmental regulation stemmed from obligations under EC law. In recent years, in response to the shortcomings of traditional methods, Italian environmental policy has undergone something of an in-depth transformation, with the introduction of new instruments replacing commitment to command and control. After reviewing the general arguments surrounding the choice of environmental instruments, the first part of this chapter explores some of the shortcomings of traditional instruments, and identifies the configuration of domestic and EC pressures animating the shift towards a new generation of environmental tools. This is followed by an examination of the advantages and disadvantages of several new tools as they have been applied in Italy. Particular attention is paid to the policy design and institutional framework necessary to achieve the simultaneous goals of increased economic efficiency and environmental protection.

Economic competitiveness and instrument choice

In the current Italian debate about sustainable development the idea frequently emerges that environmental protection could be a significant hindrance to economic growth. This is clearly a misperception that must be openly clarified. From this point of view, a useful distinction has to be introduced between regulations and economic instruments implemented in the management of environmental policy. Regulations impose a constraint on economic activity so that emissions into the environment are limited. Firms are obliged to comply with these constraints through control measures that are normally applied at the end of the pipe or through changes in the structure of the production process. They have to face a cost that is difficult to evaluate ex ante but none the less has an impact on the competitive position of the firm, and so on growth. The conventional wisdom is that environmental regulations impose significant costs, slow productivity growth and thereby hinder the ability of domestic firms to compete in international markets. This loss of competitiveness is believed to be reflected in declining exports, increasing imports and a long-term movement of manufacturing capacity abroad, particularly in pollution-intensive industries (see Golub 1998).

In the literature the use of economic instruments is normally considered more efficient whenever the marginal costs of pollution control for different firms are not the same. This means that it is cheaper to achieve an environmental goal through the use of economic instruments rather than regulations – static efficiency – and, furthermore, that the economic instruments provide a permanent incentive to improve environmental performances, while this incentive does not exist in the case of regulations since the firms have no reason to go further once the standard has been achieved – dynamic efficiency.

Even if the advantages of these new tools can be precisely defined, in practice the implementation of economic instruments to improve environmental conditions is powerfully resisted since the costs which they determine are quite noticeable and the negative impact on competitiveness – at least in the short run – can be easily perceived by the firms affected by their burden. Hence, the current view that competitiveness could be worsened by the implementation of stringent environmental policy makes it more difficult to adopt new measures, especially when global, and not domestic, environmental goods are affected by policy decisions.

In fact, this fear is misplaced, as evidence of countries deliberately resorting to low environmental standards to gain competitive advantage or to attract investments does not seem to be available. No systematic competitive impacts from disparate environmental regulations, no significant loss of markets, domestically or abroad, due to eco-dumping, no industrial migration to countries with lower environmental standards has been documented. As far as the United States is concerned, a recent study shows that:

> there is relatively little evidence to support the hypothesis that environmental regulations have had a large adverse effect on competitiveness, however that elusive term is defined. Although the long-run social costs of environmental regulations may be significant, including adverse effects on productivity, studies attempting to measure the effect of environmental regulation on net exports, overall trade flows, and plant-location decisions have produced estimates

> that are either small, statistically insignificant, or not robust to tests of model specification.
>
> (Jaffe *et al.* 1995: 157–8)

There are different reasons why the effects of environmental regulation on competitiveness are small. For all but the most heavily regulated industries the cost of complying with environmental regulation is a relatively small share of total cost of production. Even where there are substantial differences between environmental requirements within the internal market and abroad, domestic firms – and other multinationals as well – are reluctant to build less than state-of-the-art plants in foreign countries. Finally, even in developing countries where environmental standards – and certainly enforcement capabilities – are relatively weak, new plants normally embody more pollution control than is required. Therefore, 'even significant *statutory* differences in pollution control requirements between countries may not result in significant effects on plant location or other manifestations of competitiveness' (Jaffe *et al.* 1995: 158).

Recently a new view is emerging which assumes the need to use a comprehensive package of both regulations and economic instruments, and links strict environmental standards with an improvement of competitiveness in the long run. Michael E. Porter's view is now widely shared:

> It might seem that regulation of standards would be an intrusion of government into competition that undermines competitive advantage. Instead, the reverse can be true in many circumstances. Stringent standards for product performance, product safety, and environmental impact contribute to creating and upgrading competitive advantage. They pressure firms to improve quality, upgrade technology, and provide features in areas of important customer (and social) concerns.
>
> (Porter 1990: 647)

Recently, an EPA Conference also remarked that environmental regulations induce more cost-effective processes which reduce both emissions and the overall costs of doing business. Following these ideas, environmental regulations begin to be seen not only as benign in their impacts on international competitiveness, but actually as a net positive force driving private firms and the whole economy to become more competitive in international markets.

However, the previous conclusion that environmental protection could promote competitiveness is correct only when a sound environmental policy is effectively in place. In this connection it should be kept in mind that the main goal of environmental taxes is not to provide new revenues to the government, but to change behaviour. Hence, the impact on competitiveness following the use of environmental charges should be evaluated on the assumption that the rule of revenue neutrality is really accomplished, that is, also taking into account the effects deriving from the cut in other taxes made possible through the use of the revenue of environmental levies. In this case, many empirical studies either within the EC or in the United States show that any negative impact on employment or national income tends to disappear.

The use of the revenue is thus as important as the levy of the environmental charge (see Chapter 1). From this it follows that a sound environmental policy, largely utilising economic instruments, should be shaped in such a way as to take advantage of the possibility to employ both incentives and disincentives for environmental protection. The charge has the main task of internalising external effects so as to avoid a market failure and to include the real costs of the use of natural resources in the decisions taken by producers and consumers. The revenue raised through the levying of the environmental charge could then be used either to cut the rates of other taxes which have a distortionary impact on the economy or to provide incentives to smooth the conversion of firms and consumers to the new market conditions, thus overcoming the difficulties – and the consequent costs – caused by the existence of widespread market imperfections (Convery *et al.* 1996).

If this view is correctly adopted a more positive view of environmental policy emerges. The environment can be seen as a resource, an opportunity, rather than as a constraint which hinders economic growth. Hence, expenditures for protecting the environment should be evaluated not only as additional costs for industrial production, but partly as investments that in the long run could increase competitiveness through a spur to technological innovation. Environmental charges are not just a new burden for industries and consumers, but a way to cut distortionary taxes and raise welfare.

There is another important feature characterising the possible use of environmental charges. In the modern industrial society a huge change is currently under way. Science is becoming a very important production factor (Gerelli 1995) and Western industrialised countries will be able successfully to face competition from the industrialising world – where the level of wages is significantly lower – only if the possibilities linked to the use of new technologies are fully exploited and the labour force is highly qualified (Krugman 1994). This structural change requires a further integration of environmental considerations into other policy areas so that an effective model of sustainable growth might be achieved. This highlights another important task for fiscal policy, whose main burden should be lowered on labour and correspondingly raised on the use of natural resources. At the same time, the revenue from environmental charges could be used to promote innovation in the production process and to brush away the manifold obstacles that hinder the change to a post-industrial society. The main characteristic of this society is the enlargement of the advanced tertiary sector, which provides positive effects on the natural environment, energy savings, traffic volumes, generally on the functioning of public utilities and on the social and cultural structure of the society.

This is the point of view already adopted in the Delors Report (European Commission 1993), which strongly suggests a shift of the main burden of taxation systems away from labour and towards natural resources in order to promote, simultaneously, an improvement in environmental condition and increased employment. This view should be enlarged, since the whole tax system has to be screened so that most provisions with a negative impact on the environment can be removed, while those with a positive impact are strengthened. The use of the revenue must be scrutinised as well, since revenue neutrality means that the additional tax revenue could be targeted to cut the rates of distortionary taxes, to promote employment through a reduction in tax rates hitting the use of labour or to ease the road towards a post-industrial society with large beneficial side effects on the environment. As has been underlined recently by the European Commission, progress in this area is urgently needed since current tax policy in the EU

> has substantially contributed not only to maintaining distortions in the Single Market, but also – less visibly – to generating unemployment and even to creating opportunities for tax base erosion. . . . When preparing and presenting proposals the Commission will take into account the important issue of the use of taxation instruments for other Treaty objectives. The efficiency criteria for taxation can accommodate policy objectives such as health, environment, energy, transport and others deemed worth by the Treaty.
>
> (European Commission 1996)

The shift towards new instruments

In recent years a larger utilisation of environmental taxes has been on the political agenda in many European states in order, first, to overcome the limits of an environmental policy structurally based on command and control measures and, second, to provide means enabling the Treasury to cut other taxes with distortionary effects on the economy, such as income taxes with very high marginal rates or social-security contributions, mainly those levied on unskilled or low-skilled workers. This kind of manoeuvre could provide a double benefit to the economy, known in the literature as 'the double dividend' (Pearce 1991; Bovenberg and de Mooj 1994; Goulder 1995; Golub 1998).

The use of new economic instruments to replace or augment traditional command and control measures seems generally quite reasonable, especially in Italy, where past environmental policy has shown considerable limits regarding its effectiveness in guaranteeing the conservation of natural resources and the improvement of environmental conditions. Environmental legislation is anchored in the Constitution (Articles 9 and 32) – which requires safeguards for the national natural, historic and cultural heritage and recognises the right of individuals to health – and further elaborated through Acts of Parliament. The first example of environmental law concerned the provision of guidelines for air-pollution control and prevention (Law 615, 13 July 1966); it was followed ten years later by the Water Pollution Control Law, regulating the discharge of industrial and municipal effluents into surface and ground water (Law 319, 10 May 1976).

This legislation was promoted by domestic forces, having regard to the dreadful amount of polluting effluents emitted into the air and the water, mainly by the industrial system and by rapidly expanding urban traffic. But its effective implementation has been resisted by the industrial sector, with the main justification that Italy was on the way to catching up with other European countries and that Italian firms were unable to cope with the costs of extended environmental protection without impairing their external competitiveness. Public opinion, for its part, was not forcefully in favour of more effective environmental measures since it was satisfied by the continuous expansion in the level of real national income and the attendant rise in living standards of both the middle class and workers. The possible trade-off between growth and environmental protection was fully exploited by the industrial sector in order to find the political support needed to avoid the costs of pollution prevention.

At the time, discussion of the possible use of economic instruments in environmental policy was limited to a restricted group of academics (Istituto di Finanza 1970), but this policy orientation was rejected, not only by the industrial sector, but also by green groups, who were supporting the use of regulatory measures with a two-pronged argument: first, they assumed that the level of pollution prevention was uncertain when economic instruments were applied, since it was linked to the reactions of producers and consumers to price changes induced by the implementation of environmental taxes; second, the tax was rejected on ethical grounds, since it was considered equivalent to the attribution of a right to pollute.

The subsequent expansion of the overall body of environmental regulations in Italy has mainly been linked to the development of European legislation in this area. In 1982 three Decrees by the President of the Republic were issued: DPR, 8 June 1982, n.470, implementing the EC Directive on bathing water quality, allocating responsibilities to the Regions (mapping and planning) and local bodies (monitoring facilities); DPR, 3 July 1982, n.515, implementing the Directive on the quality required for surface water intended for the provision of drinking water; DPR, 10 September 1982, n.915, Waste Law, regulating waste collection, transport and disposal and implementing EC Directives 75/442, 76/403 and 78/319. Accordingly, the main body of environmental rules was established in Italy following European standards. However, this does not imply that there have been ensuing improvements in the level of environmental protection.

As a matter of fact, environmental laws formally provide a set of very stringent conditions, which represent a heavy burden for productive activity without any large positive impact on the environment, for two main reasons:

- It is very costly and largely unfeasible – due also to the dreadful conditions of public administration – to check a very large number of polluting emissions. It follows that the risk of the polluters being punished is very limited and certainly lower than the advantages flowing from not complying with the constraining normative provisions.
- If the public authority tries to force compliance with the law, the firms are able in essence to blackmail public opinion, supporting the idea that a trade-off exists between environmental protection and economic growth since environmental protection necessarily implies a cost in terms of employment and output.

If this general remark is correct, it seems that there are good reasons to implement environmental charges as a substitute for pre-existing distortionary taxes. The double-dividend argument appears sufficiently well founded, on the basis of the existing literature and specifically on empirical grounds (Majocchi 1996, 1997). In any case, it should be taken into account that the revenue from this source is substantially limited. Sweden is always – and rightly – quoted as a country that has been able to implement fiscal reform based on the lowering of marginal tax rates for income taxation, financed by a widening of the value-added tax (VAT) tax base and by new environmental taxes. But in Sweden too the total revenue flowing from the environmental charges represents only about 10 per cent of the global revenue accruing to the Treasury (Swedish Environmental Protection Agency 1997).

Even within these quantitative limits, environmental charges should be positively evaluated since, in contrast to other taxes, they do not cause new distortions, but rather:

- they remedy some distortions already existing in the economic system;
- they provide an incentive to firms to improve their productive choices, looking not only for maximum profit but for the 'optimal' level of exploitation of environmental resources as well;
- they can be filled in progressively to give firms the possibility to comply gradually with the more constraining level of environmental standards;
- they provide revenue to the Treasury which could be utilised to cut other distortionary taxes.

From a political standpoint, the switch from regulations towards the use of economic instruments has occurred only recently, and it has been inspired by domestic and European factors. In Italy there has been a radical change in the attitude of the green groups, which have largely been disappointed by the poor results following a long period of implementation of environmental regulations. This change has been further promoted by the new course of environmental policy at the European level. Particularly decisive has been the proposal put forward by the European Commission for a carbon/energy tax to curb emissions of carbon dioxide into the atmosphere, and the subsequent debate on the possibility of putting the tax revenue back into the economy in order to cut the rates of existing distortionary taxes and to promote green fiscal reform.

Within the government and the traditional political groups this idea appears largely convincing since the groups on the right appreciate the option of reducing the rates of income taxes, while on the left a cut in the level of social-security contributions charged on top of wages seems to be able effectively to promote an increase in employment opportunities. The industrial sector, however, remains generally opposed to the use of economic instruments. However, in some cases it has been possible to overcome this opposition – as in the case of the new landfill tax (discussed below) – since the problems were clearly unmanageable with the existing rules and there was considerable pressure by public opinion to find new solutions for existing environmental problems.

The remainder of this chapter explores the design and effectiveness of new environmental instruments as they have been applied in Italy. Particular attention will be paid to how revenue from new tools is used (fiscal neutrality, subsidies, etc.) and to the extent of environmental improvement expected from the implementation of the new instruments.

New instruments in practice

Revenue and incentive taxes

In Italy there is only a limited number of true environmental taxes (see Table 7.1) and these levies represent a very limited share of total revenue (see Table 7.2). As will be seen in the following paragraphs, taxes are paid at the municipal level to cover the costs of the treatment of urban wastes, while charges are used to finance collection or recycling of particular wastes. Recently a new tax has been introduced on wastes disposal through landfill and fees are paid for the use of water. These fees have recently been revised in the framework of the new rules established by the Law of 5 January 1994, n.36 (Galli Law) on the management of water resources, that provides for new charges on the complete water cycle. Some regional levies are levied also on quarries.

A fee was introduced in the past to limit the use of plastic bags, but it has been phased out and replaced by a charge on domestically produced or imported primary plastic material. This seems to be a good example of the difficulties of administrative management of an ecologically oriented charge. The excise on plastic bags was first introduced on 9 November 1988, the rate

Table 7.1 Environmental taxes and charges

Sector	*Type of tax*	*Objective*	*Basis*	*Collector*	*Payer*
Water	Waste-water tax	Partial financing of collection and treatment	Volume and quality of water	Local water company	User
	Tax on release of polluted water in the environment	Partial compensation of damage	Volume and quality of water	Local water company	Polluting enterprise
Waste[a]	Municipal solid-waste tax	Partial financing of collection and treatment	Area of housing	Commune	User
	Plastic bag fee[c] (L100 per bag)	Reducing consumption of plastic bags	Number of units on the national market	State	Producer, importer
	Charge on plastic containers and packaging for liquid	Financing of recycling	Primary material produced or imported	National consortium	Producer, importer
Oil	Charge on lubricating oil (L5 per litre)	Financing of collection and recycling	Oil sale	Consortium for used oil	Enterprises
Batteries	Charge on lead batteries	Financing of collection and recycling	Battery sale	Consortium for lead battery and lead waste	Producer, importer
Unleaded petrol[b]	Tax reduction	Promoting use of unleaded petrol	Petrol sale	State	Benefit to the consumer
Aircraft	Noise tax		Noise	State	Airline
CFC[d]	Deposit/refund	Promoting collection and recycling or disposal of CFCs	CFC volume	Consortium for CFCs	Buyer of equipment

Source: OECD (1994: 104)

Notes:

[a] Waste exports to non-OECD countries can only take place after a deposit is made

[b] There are also taxes on motor fuels and on car registration

[c] Abolished in December 1993

[d] Created in December 1993

Table 7.2 Revenue flowing from environmental, energy and car taxes (as a percentage of total tax revenue)

	1993 (%)	*1994 (%)*
Environmental taxes		
Tax on urban solid wastes	1.21	1.24
Fee on water treatment	0.25	0.25
Provincial charge for environmental protection	0.03	0.04
Contribution to the mandatory consortium for reuse and recycling	0.02	0.02
Other	0.02	0.01
Total environmental taxes	**1.54**	**1.57**
Energy and car taxes		
Tax on methane gas	1.23	1.28
Tax on electricity	0.57	0.58
Tax on petroleum gas	0.17	0.18
Municipal and provincial surtax on electricity	0.38	0.40
Car taxes	1.79	1.82
Tax on mineral oil	8.60	8.98
Total energy and car taxes	**12.74**	**13.23**

Source: ISTAT (1995: 332)

was L100 for each bag which was not biodegradable. The definition of biodegradability proved highly controversial, and has entailed first an extension of the fee to all plastic bags – disregarding the degree of biodegradability – then the mutation of the excise into a charge on virgin polyethylene films (the raw material from which the bags are produced) in order to finance the statutory consortium responsible for the recycling process. The decision to change the taxable basis of the fee was followed by the granting of a large number of exemptions: for instance, the Law of 28 December 1995 (n.549, Article 3(41)) granted a subsidy to agricultural producers who give back the residuals of used polyethylene films so that the disposal of wastes into landfill sites can be reduced. Hence, the revenue has collapsed, falling from L92.4 billion in 1993 to only L11 billion in 1995. In any case, the tax had a positive impact on consumers' behaviour, obliging them to recognise that a plastic bag has an economic value and thus favouring a more effective use of the bags, through the provision of a real incentive in favour of their re-use or the adoption of more long-lasting containers.

Regarding municipal waste management, all capital costs for the construction of municipal waste-treatment plants and landfills are currently funded by the central government. Operating costs for the collection, treatment and disposal of wastes are covered by a municipal solid-waste tax, up to now calculated according to the surface area of dwellings. But a change in this structure is now under way (see below).

Economic instruments are also used to support the recycling of plastic containers, lead batteries (with a charge, reduced on 1 November 1996 to L200 for small batteries and L1600 for large ones, allocated to the statutory consortium dealing with lead waste which recuperated a total of 153,000 tonnes in 1995) and lubricating oil (with a unit charge of L5 per litre, also allocated to the relevant consortium dealing with used oil). The objectives of these charges are twofold: first, to promote an increased saving in the use of these materials; second, to fund the functioning of the consortia charged with recycling the wastes. The same scheme has recently been adopted with the new rules implementing the EC Directive on packaging.

Six Regions have already introduced a charge on the extraction of materials from quarries. Sometimes the charge is measured according to the market value of the extracted materials; sometimes the environmental impact is taken into account when establishing the amount of the charge, following the different impact on the territory and the environment. The total amount of the charge is due to the municipalities, which are obliged to use the revenue to fund reclamations of the sites and works of environmental improvement.

The only kind of taxation that is really important in the environmental field – also from the point of view of the revenue raised – is related to energy taxes. In Italy excises are levied on all kinds of fuels except coal (see Table 7.3). Quite clearly, energy taxation has been implemented on revenue, not environmental, grounds. The main reason underlying this choice has been the very low elasticity of energy demand, which either minimises the excess burden of indirect taxation or provides a large amount of money to the Treasury. The revenue is very important from a quantitative point of view and is also relevant in an international comparison: according to a recent evaluation by the Central Institute of Statistics (ISTAT), the revenue flowing from this source – including car taxes – amounted to 13.23 per cent of total revenue in 1994 (see Table 7.2).

It should also be noted that the per capita revenue from energy taxes varies widely according to the region. It is nearly twice as high in the north than in the south (see Table 7.4). This is linked to climatic factors (heating is only significant in the north) and to existing differentials in the industrial structure (firms are mainly located in the north). Since the distributive impact of energy taxation is generally considered one of the weakest features

Table 7.3 Energy taxes in Italy

Mineral oils	*Methane*	*Liquid petroleum gas*	*Electricity*
Excise on mineral oils	State consumption tax	Excise	State consumption tax
Regional surtax on petrol[a]	Regional surtax	Value-added tax	State surtax
Value-added tax	Value-added tax		Provincial surtax[b]
			Municipal surtax[b]
			Value-added tax

Notes:
[a] In 1993 introduced only in Piemonte and Puglia
[b] From 1994 the revenue flows to the Treasury. The removal of these surtaxes is forecast in the Financial Law 1997.

of energy taxation, this point is quite important as income levels are lower in the south than in the north. In accordance with this high level of taxation, energy intensity in Italy is comparatively lower than in other industrialised countries.

While taxation of energy is normally justified by revenue purposes, tax differentiation between leaded and unleaded petrol has been introduced specifically for environmental purposes. The tax difference has varied in percentage between 6.9 per cent and 10.6 per cent, but it has been sufficient

Table 7.4 Energy taxes without VAT (billions of lire, 1993): revenue distribution among the regions

	North	*Centre*	*South*	*Italy*
Electricity	1.211	0.453	0.472	2.136
Natural gas	3.944	0.987	0.352	5.283
Mineral oils	19.390	8.268	10.007	37.765
Liquid Petroleum Gas	0.401	0.108	0.224	0.733
Total	24.946	9.916	11.055	45.917
Inhabitants ('000)	25.550	11.012	21.184	57.746
per capita revenue ('000)	0.976	0.900	0.522	0.795

Source: Majocchi (1997: 57)

to increase the market share of unleaded petrol from 9.6 per cent in 1992 to 33.8 per cent in 1994 (see Table 7.5). In this field Italian policy has followed the implementation of EC laws. However, given the good results already achieved, it seems now that such tax incentives could be more effectively utilised to curb the amount of other pollutants included in petrol, for instance benzene.

The carbon/energy tax

Following approval by the European Commission of the proposal for a Council Directive introducing a tax on carbon-dioxide (CO_2) emissions and energy (European Commission 1992), the debate in Italy about environmental taxation was fuelled by the presentation of the White Paper on Fiscal Reform by the finance minister in December 1994 (Ministero delle Finanze 1994). A few days later the government resigned. Consequently, there has been no official discussion within Parliament of this document, but none the less the White Paper represents the climax of a lengthy political and cultural debate, and can be considered an effective basis for a fruitful discussion about the future of the Italian fiscal system – even if the structure of the proposed reforms is sometimes controversial (Fossati and Giannini 1996).

This remark is particularly true when environmental taxation is directly taken into account. In the White Paper – for the first time in an official Italian document – a Pigouvian tax has been carefully considered as a mainstay of the fiscal system. In particular, the White Paper defines the three main goals of fiscal reform:

Table 7.5 Tax differentiation between leaded and unleaded petrol

Year	*Leaded petrol tax (lire/litre)*	*Unleaded petrol tax (lire/litre)*	*Difference*
1990	885.7	822.7	63.0
1991	917.9	854.8	63.1
1992	910.8	847.8	63.0
1993	942.7	853.0	89.7
1994	1019.1	911.0	108.1
1995	1097.8	989.8	108.0
1996 (January)	1111.5	1022.3	89.2

Source: Ministero dell' Ambiente (1997: 345)

- *Fiscal federalism.* In Italy, starting from the fiscal reform of 1971, there has been a substantial centralisation of the tax system, and local and regional authorities derive the majority of their funds from grants or revenue-sharing. The proposal is to shift a large amount of taxes from the centre to the periphery in order to enhance the efficiency of public expenditure and to promote the responsibility of local authorities.
- *Limits on personal taxation.* Due to the combination of a very progressive income tax and a high level of inflation in previous decades, the amount of revenue raised through direct taxation is now overwhelming. The proposal is to shift a large fraction of the tax burden away from personal income and to tax 'goods' to a greater extent. In this context, environmental taxation is clearly included.
- *Simplification.* In Italy four taxes (personal income tax, VAT, excise on petrol and the withholding tax on interests) provide, with great efficiency and little equity, some 72 per cent of total revenue. Sixteen taxes, including these four, provide some 97 per cent of total revenue. The proposal is to call a halt to the Italian 'one hundred taxes' (Tremonti and Vitaletti 1986) and to limit the complexity of the tax system, which favours extremely widespread tax evasion and represents a particularly negative feature of the current tax system.

The White Paper accepts the suggestion put forward in the Commission's proposal of introducing a carbon/energy tax with revenue amounting to about L10 trillion. This additional revenue could be put back into the economy by cutting the level of other tax rates in order to exploit a double dividend in the manner outlined by the Delors Report: the first dividend is provided by the curbing of CO_2 emissions into the atmosphere, the second by the diminution of the *deadweight loss* associated with the existence of distortionary taxes.

The White Paper embraces the idea of recycling the revenue through a cut in personal income-tax rates. Hence, households will be compensated for the increase in energy taxation, while firms will be obliged to face the increased burden of the new taxes targeted at enhanced environmental protection. The political motivation behind this choice – which explains why the European suggestion to use the revenue for cutting rates of social-security contributions has not been adopted – is probably linked to the fact that the White Paper has been prepared by the Ministry of Finance, which is not responsible for the social-security contributions and wanted to exploit the political dividend of cutting income-tax rates.

The choice that has been adopted corresponds to the general goal of shifting the burden of taxation away from persons and on to goods. In the White Paper a suggestion is put forward to cut the highest tax rate from the current level of 51 per cent to 40 per cent or 45 per cent, according to different options, and to reduce the average rate for the majority of taxpayers to below 25–30 per cent.

A revision of energy taxes is envisaged, following the general ideas prevailing in the European debate, according to the carbon content of each energy source, while ensuring that an effort is made to keep the relative tax burden between the different sectors – households, industry, transport – generally constant. In line with the overall inspiration of the fiscal-reform project, existing energy taxes are included in a General Energy Tax that applies the same regime to all the different energy sources and is strictly coordinated with VAT.

The tax increase will be applied to the industrial sector only if a decision is taken at the European level to implement a common carbon/energy tax. This provision is considered unavoidable since a unilateral implementation of the carbon/energy tax will significantly weaken the overall competitiveness of Italian industry. Increasing existing energy-tax rates would provide an additional revenue amounting to L10 trillion according to two different options that are considered in the White Paper, each of which calls for an increase in the price of energy of about US$7 for a barrel of oil equivalent, largely similar to the European proposal, which results in a price increase of US$10 per barrel of oil.

The economic impact of this manoeuvre is different in the two scenarios (Majocchi 1997). In the first case all the energy sources and all the sectors – households, industry and transport – would be charged the new energy levy, shaped according to the carbon content of each energy source. In this case it is assumed that an agreement has been achieved at the European level for a harmonised implementation of the new energy tax. Furthermore, in the White Paper the new tax burden is modulated in a way that leaves substantially unmodified the relative prices in the different sectors and is equivalent to a price increase of L126 per litre of petrol and L142 per litre of diesel.

In the first year following the implementation of the manoeuvre – and taking into account the impact of the cut in income-tax rates – GDP increases by 0.56 per cent, while in subsequent years it remains more or less stable. The consumption-price level increases in the first year (+0.75 per cent), but then the rate of growth diminishes rapidly, while the price impact on production costs in the industrial sector is more sensible in the first year (+1 per cent), but in the following years it tends rapidly towards zero.

In the second scenario, where it is assumed that no harmonisation takes place at the European level, industry is not charged and the new tax burden – which is shaped according to the carbon content of each source – is concentrated upon households and transport, while the price changes following the manoeuvre are not proportional to the previous prices in the different sectors. In this case the price of petrol increases by L161 per litre, while the price of diesel increases by L182 per litre.

The impact on consumption prices is very similar, while it is naturally different as far as industrial prices are concerned since industry is exempted in this case. But it is important to remark that the rate of growth of GDP is practically the same as in the first hypothesis. Exemption of the industrial sector has no real impact on the growth of the economy.

These are the main economic changes that would follow from the implementation of this proposal of fiscal reform regarding the increase in energy taxes, whose revenue is targeted to a cut in income-tax rates. The most important effect of this manoeuvre, however, will concern the environment, since it provides a very relevant impact on air quality. The combined index of air quality – including CO_2 emissions, sulphur dioxide (SO_2), nitrogen oxides (NO_x and particulates) – declines from 97.1 in 1994 to 95.4 in the year 2000 in the reference scenario, and to 92.6 taking into account the impact of the manoeuvre, while the rate of growth in the CO_2 emissions drops from 110 in the reference scenario to 106.25 in the year 2000 (100 being the index level in 1990).

Recent waste and water taxes

Up to now there has been no specific follow-up to the proposals included in the White Paper. In any event, however, they have yielded one important result: the debate about the use of environmental taxes has been refuelled, and the limits of an exclusive use of regulations in the field of environmental policy have largely been recognised. As a result of this renewed awareness that a more balanced mix of regulatory and economic instruments is needed in order to achieve a more effective policy, the Ministry of the Environment has been able to suggest the inclusion of the proposal for a new landfill tax in the draft budget for 1996, which was approved by parliament in December 1995.

The goal of the tax is to eliminate one of the justifications for the adoption of this kind of waste disposal, which seems the most noxious to the environment. In particular, the Ministry of the Environment had identified two central problems with the current arrangements in the area of waste management: the costs of waste disposal faced by the individuals are not related to the amount of waste they discard, nor do they reflect the true economic or environmental costs of the waste disposal method; the lack of relationship between the waste produced and the price charged for collection and disposal represents a clear failure with respect to the implementation of the polluter-pays principle.

Landfill has traditionally been the favoured option for disposal in the EU. Five member states, including Italy, currently rely on landfill for more than 80 per cent of their waste disposal. One solution becoming increasingly popular to halt this dependency is the landfill tax. By adding to the costs of landfill, the tax modifies the relative prices of different waste-management options, creating a disincentive to landfill relative to other waste-disposal options, thus favouring a shift in waste management one step up the hierarchy, i.e. toward incineration with energy recovery. But even if this is a step in the right direction, it must be underlined that a landfill tax on its own is unlikely to achieve an increase in the level of recycling.

The landfill tax was introduced in Italy by the Law of 28 December 1995

(549, Article 3 (24–40)). The goal of these provisions is to favour lower production of wastes at the source and recovery of energy and raw materials from wastes. From 1 January 1996 landfill-site operators must pay the tax to the Regions and will therefore raise their charges to waste disposers. The taxable base is represented by the weight of wastes, and the tax rate is differentiated according to the environmental impact of different kinds of wastes. The rate for inert wastes can be chosen by the Regions in the range of L2–20 per kilogram; for special wastes L10–20, for urban and assimilated wastes L20–50.

The expected revenue for the Regions in 1996 amounted to over 1 trillion lire. One fifth of it must be devoted to a special regional fund that could provide incentives to favour lower production of wastes, to recover energy and raw materials from wastes, to reclaim contaminated sites and to finance the creation and conservation of natural protected areas. The original idea was that a larger share of the revenue should be used to fund environmental improvements; but this possibility has been excluded due to the pressure by the Treasury trying hard to cut the amount of the overall deficit in order to comply with the financial constraints established by the Maastricht Treaty.

One of the risks linked to the adoption of a landfill tax is the promotion of fly-tipping. In order to avoid the illegal disposal of wastes, the law establishes a shared responsibility – in the payment of the tax and the related fines – of waste disposers and the user and/or the owner of the land where the disposal takes place, while keeping unchanged the criminal sanctions against this type of activity.

This tax was conceived as an important instrument to change relative prices of different kinds of waste disposal. The important point is that the tax must be seen as only a part of a general strategy to reduce the amount of wastes going to final disposal without any treatment. The decision to implement new economic instruments in this field has been backed up by the promotion of voluntary agreements with important industrial sectors – mainly cement and electricity production – to use refuse-derived fuel (RDF) as an alternative fuel. The use of this instrument – the carrot – has been successful since it was supplemented by the implementation of the tax – the stick.

This change in policy has led to the adoption by the government of new legislation on wastes which transfers into Italian law all the previous Directives approved by the European Union (91/156, 91/689, 94/62). One important point of this new legislation, (D. Lgs. 5 February 1997, n.22) is the adoption (Article 49) of a two-part tariff for the collection and treatment of urban wastes, in substitution for the pre-existing tax on municipal solid wastes: the first part of the tariff is targeted to cover the investment costs needed to provide the waste treatment; the second part is proportional to the amount of wastes produced and to the operational costs of treatment.

The idea is to promote a reduction at the source of the amount of wastes produced and, at the same time, to support the industrial sector which should necessarily emerge in order to achieve the ambitious recycling target established by the new legislation. In particular, the six-year target for municipal solid wastes is an amount equal to 35 per cent of separate collection (Article 24); for packaging, the target for the subsequent five years (by weight) should be reached (Article 37): re-use as raw materials or for energy recovery between 50 and 65 per cent, recycle between 25 and 45 per cent, with a minimum established for each packaging material.

Finally, according to the provisions defined by the Law of 5 January 1994 (n.36; Galli Law on the management of water resources), a new tariff has been established for the use of water resources (Article 13). The new regulation of water tariffs was recently approved by the Italian Ministry of Public Works (Decree, 1 August 1996, published in the *Official Journal* on 16 October 1996). This new regulation provides a sort of benchmark to define the composition of total costs and water charges to be implemented. It is supposed to lead both to an increase in the prices paid by consumers and to efficiency gains as a 'price-cap' is also gradually introduced.

The new regulation should also break with a past where tariffs were mainly set according to macroeconomic and distributional aims – in order to keep water charges very low compared with those prevailing in the other countries of the European Union – and cost issues played a minor role. The composition of the new tariff is supposed to be defined following the decree of 9 April 1991 (n.127), which, in turn, is based on EU Directives 78/660 and 83/149.[1] The new legislation is to be revised after the incorporation into Italian law of Directive 91/271. The opportunity cost of water use is not considered, but this is firmly in the Italian tradition of price regulation made by lawyers and engineers without intervention by economic analysts.

Promoting green tourism

Some interesting ideas are currently floating around about the possibility of using economic instruments to internalise the external costs generated by tourism and to promote sustainable tourism. Due to the existence of these external costs, tourism demand is too high since prices do not reflect true social costs. At the same time, it must be remarked that the benefits of tourism flow largely to the private sector, while the costs for the conservation of the artistic and natural estate are mainly borne by the public sector.

The concept of carrying capacity is highly relevant in planning for sustainable tourism development. It determines the limits of development and visitor use of an area without degrading its environmental quality or the visitor experience, and helps to establish the optimum use of tourism resources. Carrying capacity, in the general philosophy of a win–win strategy, is supposed to combine economic as well as environmental arguments. The aim is to find the optimal combination of both interests. This may lead to a compromise whereby the best solution is selected from neither an environmental nor an economic viewpoint (DRI 1994).

In Italy 35.7 per cent of total tourism is represented by international tourism, which is evenly divided between tourism from the European Union and tourism from abroad. Italy suffers particularly from pressures on the urban environment caused by tourism. The implementation of a rigorous carrying-capacity policy in historic cities will have a positive impact on their future development. From this perspective it also seems relevant that pricing devices could be implemented with a twofold goal: to limit the level of demand and to provide financial means for conserving Italy's historic and artistic estate.

For many years Italy has implemented a form of taxation on tourism through the visitors' tax, paid by people visiting a tourist site and set according to the number of nights spent in hotels. However, this tax was cancelled in 1989 in order to promote tourism (disregarding its environmental impact) and is currently implemented only on a regional basis in Trentino-South Tyrol. The limit of this tax is that it hits only one kind of tourist service. A discussion is currently under way about the possibility of reintroducing this kind of taxation in the Italian tax system. A differentiation in the tax rates according to the different prices paid in hotels could be considered for equity purposes.

A proposal along these lines was included in the Draft Financial Law for 1997. Article 74 foresees that municipalities would have the option to levy a tax on non-residents in the area who spend a night in a hotel or other residential facility – in this case only if the house has been rented through an agency, in order to guarantee effective oversight and enforcement of the tax provisions. The tax could be varied according to the price of accommodation, with a maximum limit of 5 per cent. However, the possibility of using the visitors' tax to promote sustainable tourism was rejected during the debate within Parliament, mainly for political reasons, since the Financial Law was already establishing a large increase in the amount of overall taxation and the government did not want to be characterised as excessively taxation-prone. Hence, the topic is still on the political agenda and many Italian historic towns seem interested in adopting this kind of environmental tax.

A different solution is now under discussion in the Trentino-South Tyrol region. The idea is to bring in two different types of taxes:

- A tax on tourist services to be paid not only by the hotels but also by the owners of other activities whose earnings are affected by tourism.[2]
- visitors' tax to be paid by people who use secondary houses in a municipality different from their own. Taxation of secondary houses is currently accomplished by increasing by one-third the revenue flowing to the owner. However, since the income tax is levied on the average (normal) revenue, the disincentive effect on the ownership of secondary houses deriving from this provision is largely limited.

Conclusions

In conclusion, there has been a certain shift in Italian environmental policy towards a greater use of economic instruments. This change can be explained by two different pressures:

- the costs of the extensive but largely inefficient body of current regulations, which put a huge burden on the Italian industrial structure without significant improvements in the state of the environment;
- the model of EU legislation, where a greater use of economic instruments is forecast.

This does not mean that this extension in the use of economic instruments is uncontroversial. The industrial sector is more opposed to new taxes and charges than to regulations, a position whose main justification lies in the assumed negative impact of environmental taxes on competitiveness. However, a more positive view is now emerging, especially in the context of a greening of the tax system (Moret and Ernst & Young 1996), as suggested in the White Paper on Fiscal Reform. This is probably the main point emerging from an overview of the Italian experience. The level of taxation is currently considered too high and it is quite difficult to suggest the implementation of new taxes, even for environmental purposes. A larger use of economic instruments is conceivable only if the revenue flowing from these new taxes and charges – targeted at improving environmental conditions – is used to cut the rates of pre-existing distortionary taxes. This means that the underlying philosophy must be similar to the one adopted by the European Commission's proposal for a carbon/energy tax or to the idea put forward in the Delors Report: a cut in the level of taxes levied on labour, funded by the revenue from environmental taxes.

Notes

1 The benchmark tariff in the new legislation has the following structure:

$$T_n = (C + A + R)_{n-1} * (1 + \Pi + K)$$

It is actually based on the notion of Full Cost Recovery, as C is operation costs, A represents capital depreciation and R represents the return on capital investments. Price increases are limited by Π, the rate of inflation, and by K, the price cap, which, after the first year of implementation of the new legislation, will vary between 5 per cent and 10 per cent.

2 The tax base is represented by 100 per cent of the turnover defined according to VAT rules for operators acting directly in the tourist sector, while the share of the turnover on which the tax is levied will be lower for other operators involved in activities that are not totally dependent on the tourist sector. The tax administration could be managed according to the rules defining the VAT tax base and the taxpayers could recoup their tax duties at the same time as they fill in their VAT form.

References

Bovenberg, A. L. and de Mooj, R. A. (1994) 'Environmental levies and distortionary taxation', *American Economic Review* 4: 1085–9.

Convery, F., Majocchi, A. and Pearce, D. (1996) 'Economic incentives and disincentives for environmental protection', background document for the European Commission/Italian Presidency Conference, Palazzo Altieri, Rome, 7 June.

DRI (1994) *Potential Benefits of Integration of Environmental and Economic Policies: An Incentive-based Approach to Policy Integration*, report prepared for the European Commission (London: Graham & Trotman).

European Commission (1992) *Proposal for a Council Directive Introducing a Tax on Carbon Dioxide Emissions and Energy*, COM(92) 226 final (Brussels).

—— (1993) *Growth, Competitiveness and Unemployment* (Brussels).

—— (1996) *Taxation in the European Union* (Brussels), 20 March.

Fossati, A. and Giannini, S. (1996) *I Nuovi Sistemi Tributari* (Milan: Franco Angeli).

Gerelli, E. (1995) *Società post-industriale e ambiente* (Bari: Laterza).

Golub, J. (ed.) (1998) *Global Competition and EU Environmental Policy* (London: Routledge).

Goulder, L. H. (1995) 'Environmental taxation and the double dividend: a reader's guide', *International Tax and Public Finance* 2: 157–83.

ISTAT (1995) *Rapporto Annuale 1995* (Rome: ISTAT).

Istituto di Finanza (1970) 'Problemi economici per la difesa della natura', *La Difesa della Natura: Aspetti Economici, Urbanistici, Giuridici* (Pavia: Istituto di Finanza).

Jaffe, A. B., Peterson, S. R. and Portney, P. R. (1995) 'Environmental regulation and the competitiveness of U.S. manufacturing: what does the evidence tell us?', *Journal of Economic Literature* 33: 137–63.

Krugman, P. (1994) 'Does third world growth hurt first world prosperity ?', *Harvard Business Review*, July–August: 113–21.

Majocchi, A. (1996) 'Green fiscal reform and employment: a survey', *Environmental and Resource Economics*, December: 375–97.

—— (1997) 'Environmental taxes in the Italian White Paper on Fiscal Reform', in C. Jeanrenaud (ed.) *Environmental Policy Between Regulation and Market* (Basel: Birkhäuser).

Ministero delle Finanze (1994) *Libro Bianco sulla Riforma Tributaria*, (Rome).

Moret and Ernst & Young (1996) *Tax Provisions with a Potential Impact on Environmental Protection* (Brussels), September.

OECD (1994) *Environmental Performance Reviews: Italy* (Paris: OECD).

Pearce, D. W. (1991) 'The role of carbon taxes in adjusting to global warming', *Economic Journal* 101: 938–48.

Porter, M. E. (1990) *The Competitive Advantage of Nations* (New York: Free Press).

Swedish Environmental Protection Agency (1997) *Environmental Taxes in Sweden. Economic Instruments of Environmental Policy* (Stockholm).

Tremonti, G. and G. Vitaletti (1986) *Le Cento Tasse degli Italiani* (Bologna: Il Mulino).

PART II

NEW INSTRUMENTS AT THE EU LEVEL

8

NEGOTIATED AGREEMENTS IN EU ENVIRONMENTAL POLICY

Jan Willem Biekart

Introduction

This chapter addresses the question whether and how negotiated agreements between government and industry can improve the environmental performance of companies. In answering this question, as the representative of an environmental non-governmental organisation (NGO),[1] the author wants to illustrate four propositions:

- negotiated agreements in themselves are not more than a minor step towards sustainable production;
- negotiated agreements will only work if there is an effective regulatory framework with regard to performance standards, so that, at a minimum level, formal equality between companies is assured and free riders can be dealt with;
- public access to information on the results of agreements and the performance of individual companies is an essential requirement;
- though negotiated agreements offer both opportunities and threats, as with any change, the balance of judgement depends on the credibility of agreements – credibility requires strong commitment of the parties to the targets of an agreement and the willingness to take the necessary actions whenever troubles arise in the implementation process.

The scope of this chapter is limited to negotiated environmental agreements between government and the secondary sector (i.e. industry), though agreements with other economic sectors are also known, as well as agreements in which no (national) public body participates (like neighbourhood agreements and self-imposed obligations of industry). The geographic scope is

Europe, but most examples are from the Netherlands. In order to obtain the necessary depth, our subject is put into the context of a much more challenging issue: how can the environmental performance of European industry be improved substantially through policies of the European Union or its member states.

The next section identifies some trends in environmental thinking in industry and the challenges that lie ahead. The subsequent section discusses negotiated agreements in the context of the possibilities for the improvement of European environmental policy as a whole. Next, the author elaborates on the 'rules of the game' for negotiated agreements, based in particular on the many Dutch experiences, but with an eye towards experiences in other member states. Then, some often heard questions regarding negotiated agreements are answered, while the final section draws conclusions about the value of negotiated agreements in EU environmental policy.

Environmental thinking in industry: issues

This section addresses the issues in environmental thinking that have emerged in influential bodies of industry over the past few years. A focal point of industry thinking was the United Nations world environment conference in Rio de Janeiro in 1992. Bodies of industry which have since become active in promoting voluntary agreements are the International Chamber of Commerce (see, for example, WICE 1994), UNICE (European employers' organisation), and many transnational companies and influential trade organisations like CEFIC (European chemical industry).

Public authorities in many industrialised countries are reconsidering their role and ambitions. In particular, cooperation between policy-makers and industry has become a trendy issue. In EU environmental policy a change of course was laid down in the Fifth Environmental Action Programme of 1992, which emphasised the responsibilities of industry. In 1995 the European Commission actively started exploring the opportunities for negotiated (or voluntary) agreements. Many industrial countries within and outside the EU (like the United States, Japan and Korea) were already acquainted with these instruments in some form.

However, reaching sustainability in industrial production is seldom at the core of the debate on new instruments in environmental policy. Rather, central issues in the worldwide debate on cooperation between governments and industry in environmental matters are as follows:

- increasing use of voluntary or negotiated measures;
- more flexibility in policy implementation in combination with consistent policies and clear, long-term targets;
- attention to cost-effective policies and priority-setting, stimulating an integrated approach towards environmental problems.

These are the recurring issues in industry's policy papers, memoranda, etc. to governments, often implying the need for fewer regulations. This is, in essence, an economic and not an environmental agenda. The following is a characteristic quote: 'Voluntary agreements offer the best guarantee to control the influence of the chlorine industry on the environment,' said former CEFIC president Daniel Janssen at a conference of Euro-Chlor in Brussels (NCI 1995: 1). The same kind of statements could be heard at a UNICE workshop in March 1995 (*Europe Environment* 1995: 1).

Industry representatives, but also some governments and politicians, tend to deny or conveniently forget the hard fact that every survey on industry's motives for doing something about environmental problems comes to the same conclusions: for the vast majority of industries, regardless of their country of origin, a corporate environmental policy depends on the presence of legislation. Moreover, in companies which aim to go beyond compliance with legal requirements, the main driving force is gaining advantages by anticipating legislation.

While the environment now occupies a permanent position on the managerial agendas of most bigger industries, business has responded to the challenge in its own, well-known way: reducing it to a management problem of limited proportions and seeking solutions in rational and proactive management. Hence the relative popularity of environmental management systems and the scarcity of business initiatives which explore more far-reaching solutions that translate the concept of sustainability into concrete actions.

Is industry really going green or is it all 'greenwash', as Greenpeace stated in a 1992 report (Greenpeace International 1992)? The credibility of industry's environmental performance is one of the central issues in this debate between industry and environmentalists. Differences between individual companies on this point have grown widely, however, which makes generalisations more difficult. There are enormous divergences in attitudes towards environmental issues between branches of industry, between companies and even within companies. There are also growing market niches for environmentally sound products. An understanding of this is essential in the discussion on new instruments in environmental and industrial policy.

The accountability of industry's environmental performance is still extremely low. Only a very small part of industry (less than 1 per cent) is publishing annual environmental reports, of which only a few are externally verified. Accountability is only partial; for example, products and raw material input are usually excluded (UNEP 1994). Implementation of public release and transfer registers (PRTRs) by governments, documenting emissions of all industrial companies, are strongly resisted by industry organisations and individual companies. At the EU level, the Pollution Emission Register (PER) initiative of the Commission, for instance, is so strongly opposed by CEFIC that all work on it seems to have stopped. This represents a remarkable lobbying result by the chemical industry, because transparency of performance is a prerequisite of successful negotiated agreements. In the United States the existing Toxic Release Inventory is under pressure from certain parts of industry and several politicians, in order to reduce the extent of its right-to-know impact.[2]

Still, there is value in the concept of shared responsibility between governments and industry, because it seems unlikely that sustainability can be reached solely through command and control. The enormous reservoir of industry's creativity must be used and refocused on issues of sustainability.[3] However, to overcome a number of major dilemmas it is indispensable that governments dare to make difficult decisions, sometimes against established interests.

The vehicle for realising sustainable development must be a credible environmental policy. Attention in the discussion on new instruments in EU environmental policy is too much distracted by claims that command and control and self-regulation are mutually exclusive; that environmental policy must be comprised entirely of one rather than the other (as is noted in Chapter 1, similar polarisation plagues the free-market environmentalism debate). This is in fact a false dichotomy. A credible environmental policy means that transparency, effectiveness, equity and efficiency are criteria of the highest order. In practice, this will have to lead to an intelligent mix of different instruments: regulatory, economic (like subsidies and ecotaxes), social (like negotiated agreements), fiscal (like special tax schemes for green stocks) and information instruments (like education and technology development and transfer programmes), giving front-runner companies some advantages and treating laggards with a straight command and control approach.

Policy context: the need for improved environmental performance from industry

The already limited advantages of negotiated agreements will only become evident in a credible policy which is designed to put the right responsibilities on the right shoulders. Otherwise negotiated agreements become an isolated and useless or even dangerous instrument which undermines existing policy instruments. This section devotes some attention to this wide-ranging issue.

The sense or otherwise of negotiated environmental agreements should be analysed in the context of a number of logical requirements of effective environmental policy:

1 It enforces transparency and accountability.
2 It ensures that market forces work for you.
3 It provokes creativity.

4 It creates flexibility.
5 It exhibits coherent design.

First, improving the environmental performance of industry demands transparency and accountability. There is a great need to draw a company's performance out of the shadows of uncontrollable claims or even complete silence. The history of the American Toxic Release Inventory has shown that dramatic improvements in a company's performance can be achieved when the company is given the possibility of competing in the market or in public relations on the basis of a credible environmental performance. It provides an opportunity for shareholders, banks, insurance companies and business partners to obtain answers to their questions regarding the environmental risks of companies and their products, thereby improving the credibility of both government and industry in society. Therefore, vital instruments are:

- a legal obligation for companies to submit their emission data to a central, publicly accessible register, or PRTR (Pollutant Release and Transfer Register; for example the EU Pollution Emissions Register, if it does not languish) – in 1996 the OECD developed guidelines for PRTRs as a consequence of the decisions at the UN conference in Rio in 1992 (OECD 1996);
- a legal obligation to publish the environmental characteristics of chemical substances for which a company requests market access;
- a legal obligation for industrial companies to publish annually an adequate environmental report which is externally verified (similar, for instance, to the reporting requirements of the US Security and Exchange Commission);
- public access to all information pertaining to negotiated environmental agreements and their results, including information at the level of individual companies, as well as evaluations of the enforcement of environmental regulations.

Second, as the market is a powerful force in creating effective regulation with regard to the environment, one must look for policy instruments which change the rules of the economic game so that the right competition impulses are given and differences between the environmental performance of individual companies can play a clear role in the marketplace. Instruments which might be expected to be highly effective are:

- the introduction of extended producer responsibility, i.e. the producer of products remains responsible for them during their user and end-of-life phases (Greenpeace International 1995a);
- the introduction of green taxes (Greenpeace International 1995b; Dietz *et al.* 1995);

- a shift from labour taxes to taxing the use of the environment;
- strong legislation on liability for environmental and health damage, including access to justice;
- a high-quality certification scheme for environmental management systems (towards which EMAS is a very first step) (see Chapter 10);
- an ambitious ecolabelling scheme (see Chapter 9).

Third, while it is generally believed that environmental regulations usually do not stimulate innovation and creativity, this is not an inevitable characteristic of regulations, let alone of environmental policy instruments in general. Stimulating constructive creativity in industry in order to reach better solutions for environmental problems is an invaluable quality of environmental policy. To achieve this:

- set challenging, longer-term environmental targets with a commitment by industry to reach them;
- whenever possible, do not prescribe the application of a specific pollution-abatement method, but set a performance goal and let industry find how best to achieve it;
- stimulate cooperation between companies located on a particular site or within a limited region by analysing whether waste streams of one company (solid or fluid waste, waste heat, waste water) can be used as inputs for another;
- look for coalitions with companies who wish to go (much) further than 'business as usual' and find ways of rewarding them if they indeed perform better.

Fourth, flexibility is a valuable aspect of a more effective environmental policy. In contrast to creativity, which is essential for meeting longer-term and challenging objectives, flexibility is more focused on the shorter term. For example, industry should be allowed to discuss with authorities the setting of environmental priorities on the basis of cost-effectiveness and investment realities, as long as the objectives themselves, fixed by these authorities, will not be violated and will be met on time.

Fifth, coherence in the design of the legal framework is essential, though this is quite difficult to achieve. It focuses on the elimination of bloated and burdensome bureaucracy, the coordination of legislative scope, definitions and terminology, as well as on sufficient financial resources for manpower and training in order to implement and enforce environmental regulations.

If we compare the characteristics of negotiated agreements with these requirements, agreements might score well on several points. However, the opposite is just as likely to be the case, depending on the way the agreements are negotiated, designed, implemented and enforced. The next section will therefore focus on the proper rules of the game for negotiated agreements.

The rules of the game

In this section we would like to communicate experiences with negotiated agreements, particularly in the Netherlands. Based on many experiences with environmental agreements, especially with the many mistakes made in the first generation of agreements in the late 1980s, a number of 'rules of the game' are derived (for the Dutch literature on these rules, see Algemene Rekenkamer 1995; Winsemius 1993; Bogie 1993; Commissie voor de Toetsing van Wetgevingsvraagstukken 1992). We want to add that in the Netherlands guidelines for agreements in general (not just environmental ones) have been formalised and published by the government (Ministerie van Algemene Zakan 1995). These guidelines have no formal status.

Our 'rules of the game' are as follows:

1 *Ambitions criteria*:

- The environmental targets of the agreement must be set high enough so that companies perform better than would be the case with 'business as usual' or mere compliance with existing legislation. In general, this means that the targets are not negotiable by industry, but rather are based on a long-term environmental policy devised by the government.
- Wherever possible, the objective must be to award companies that do more than is minimally required (compliance, 'business as usual'), for example through flexibility in reaching specified targets or certain financial benefits, and to prevent free riders by withholding from them these rewards and by confronting them with a straightforward command and control approach. This objective becomes difficult in a situation without regulation or other 'standards'.

2 *Choice criteria*:

- The agreement must not violate existing legislation.
- The characteristics of the parties involved are crucial for a successful agreement. These characteristics pertain to knowledge, professionalism, representativeness, negotiating position, level of organisation, credibility and mutual respect.
- Parties must show a strong commitment to the environmental objectives of the agreement and be prepared to solve effectively troubles arising in the implementation process.
- In general, a limited number of parties (companies, authorities) must be involved.
- The choice of the instrument must be considered according to the situation and in the context of the implementation of other and

additional policy instruments. A regulatory framework for the issues agreed upon, which defines minimum performance standards for industry, is always necessary and should therefore be in place or at least be prepared at an early stage for eventual implementation.

3 *Design criteria*:

- The agreement must offer a clear solution (quantified targets) for a well-defined problem achievable in steps (staged approach) within a specific period of time.
- The agreement must make clear who the parties to the agreement are and what their obligations are.
- The agreement must offer safeguards for third parties, including publication of the text, public access to periodic monitoring results and to the contribution of individual companies, and verification of the results.
- The agreement must tackle the free-rider issue through the possibility of sanctions and must be binding on all parties.
- The agreement must offer some kind of reward for companies going further than compliance and/or 'business as usual'.
- Procedures for consultation of the parties and for changing or terminating the agreement must be defined.
- The results must be evaluated and made public, and the procedure for non-compliance with the targets of the agreement must be foreseen.

We will illustrate the meaning and relevance of these rules by elaborating on a range of Dutch examples of integral and single-issue agreements, respectively.

Experiences with 'integral' agreements

Experiences with *integral agreements*, i.e. agreements which establish medium to long-term targets and cover a whole set of environmental parameters of a particular branch of trade, are very limited within the EU (as well as outside it). The Netherlands stands out clearly with its target-group policy on industry, which has received widespread attention throughout the industrialised world. This, however, is not the only type of negotiated agreement in the Netherlands. Of the approximately eighty agreements with industry, only ten are integral agreements (Öko Institut *et al.* 1996). Together, these agreements (with one exception) address fewer than 300 companies, which emit the majority of industrial pollution in the Netherlands.

Integral negotiated agreements are both a communicative and a management instrument, in contrast to most single-issue negotiated agreements. The

Dutch integral agreements build on existing command and control-type instruments, but strive to incorporate the commitment of industry to long-term objectives, a real integrated approach, flexibility for industry, and transparency for the parties and the general public, which cannot be stressed enough.

In most cases it is still too early to assess the effectiveness of these integral agreements, but in some cases enough results are available to draw meaningful preliminary conclusions. Our experience is that negotiated agreements do not work, or have at least a high risk of failure, unless they meet the rules described above (see also Biekart 1995).

Chemicals

The best example of (relative) success has been the negotiated agreement with the chemical industry. In April 1993 the chemical industry – not only the trade organisation but also most of the 130 individual companies – signed an agreement with the Ministries of Environment, Economic Affairs and Water Management, the twelve provinces and the organisation of municipalities. It contains quantitative targets for the main emissions (air, water, soil, waste, energy) for 1995 and 2000, with indicative targets for 2010. For most emissions, reductions between 50 per cent and 99 per cent relative to the year 1985 are required. These targets are reasonably ambitious and non-negotiable. They do not bear the character of standards. The reduction objectives were derived directly from the National Environmental Policy Plan (NEPP), which was adopted by parliament in 1989. For its part, the NEPP was based on a quantitative assessment of the environmental situation in the Netherlands, entitled *Concern for Tomorrow* (RIVM 1988). The agreement is a declaration of intent which creates obligations, but no liabilities.

Under the agreement, companies are obliged to write a company environmental plan (CEP), in which they indicate how they want to implement the objectives of the agreement in terms of concrete measures. This is done in steps – a CEP covers a four-year period and must contain an outlook for the next four-year period. Every four years the CEP is completely revised and every year a report is compiled on the actual progress made. Thus, in the time-span of the agreement four or five CEPs will be written by each participating company. An essential article in the agreement is that companies should implement the agreement by applying 'state of the art' technologies, which means technologies considered to be proven and which do not entail excessive costs. This is anyway a requirement in Dutch permit procedures and many such technologies have been described in guidelines for permit giving authorities. If the environmental targets of an agreement cannot be reached by these means, then best available techniques must be considered. Companies must justify the choices made in their CEP with arguments. If the CEP is considered acceptable by the permit-granting authority, it is taken as the basis for the legally required periodic revision of permits or for a new permit.

Herein lies one of the chances for awarding participating companies more flexibility. Industry itself is not so much opposed to ambitious objectives, as long as it has enough time to anticipate them. Industry favours highly certainty and stability in what is being asked of it. Through the CEP a company gets the chance to anticipate long-term objectives and enjoys the liberty to propose whatever measures it wishes to take first, taking into account considerations like cost-effectiveness, depreciation of existing installations, etc. Also, negotiations take place with regard to an integral package of measures, while formerly the company had to talk to different permit-granting authorities for the different environmental aspects of its production site. This also means that under the agreement it can be considered acceptable that a company does not yet implement BATNEEC for water if it is logical to give high priority to BATNEEC for air emissions (to put it simply).

The CEPs are available to the public on request, just like the yearly evaluation reports. This is particularly important, because in this way the environmental performance and plans of a company can be monitored by third parties, which is a big stimulus in the dynamics of the agreement. The progress of the sector as a whole is monitored by adding up the results of the individual CEPs and comparing these with the objectives of the agreement. The results are evaluated by a formal body, the Overleggroep Chemische Industrie (Chemical Industry Deliberation Group). This body is responsible for the correct implementation of the agreement and consists of representatives of the chemical industry and several public authorities. Bottlenecks in the evaluation study are traced to the individual company, where the possibilities of specific action are discussed and (if agreement is found) measures are taken. Usually this will mean applying measures beyond BATNEEC. If no agreement can be found, there is always the permit procedure to fall back upon (with its public inquiry and appeal procedure); in this procedure the authority can ask of the company what it thinks is necessary. Of course, the company (like third parties) can also object to these permit prescriptions and can go to court. The permit procedure with its legal requirements thus forms the safety net.

In mid-1996 the results of the first round of chemical industry CEPs were compiled and bottlenecks have been identified. It is clear that most of the year 2000 targets will probably be reached, but some 20 per cent of the targets will not, at least not on the basis of the first generation of CEPs. Most of these bottlenecks have an economic background (too expensive to solve in the view of companies), and not a technical one. A substantial number of companies contribute to three of these bottlenecks: nitrogen oxides (NO_x), vinylchloride and carbon monoxide (CO). The other problems lie, generally, with only one or a few companies.

Several evaluations of the first generation of CEPs have made it clear that industry does not actually do more than was already foreseen in permits or other arrangements with authorities (Inspectie Milieuhygiëne 1995; Ministerie van Verkeer en Waterstaat 1995; Inter Provinciaal Overleg

1996). In fact, emissions into water have probably decreased less quickly than would have been the case without the agreement. Emissions into air, on the other hand, have probably been reduced more than in a scenario without the agreement. A possible reason for this is that the integral approach – examining all environmental problems of a site – made clear that the reduction of emissions into air was relatively more urgent than reduction of emissions into water. This, in turn, has a range of causes, which cannot be discussed here. The level of strategic environmental thinking in the chemical industry still proves to be generally low, which is a subject of concern and action by the trade organisation VNCI (Society of the Dutch Chemical Industry) and others. Still, the CEP instrument has proved to be a big stimulus for companies to think integrally about the environment. Furthermore, through the writing of a CEP many companies have been confronted with the fact that they still cannot quickly generate accurate environmental data. This shows the urgency of the introduction of good environmental management systems. The extent to which companies have really been given more flexibility in the case of good behaviour or have been confronted with strong permit requirements in the case of bad behaviour has been subject to an interesting evaluation published by the Environmental Inspectorate (Inspectie Milieuhygiëne 1997).

Oil and gas

Apart from this case in the chemical industry, some integral negotiated agreements specify inappropriate objectives and suffer from poor design. The June 1995 agreement with the oil and gas exploration and production sector, for example, has a more or less similar design to the agreement with the chemical industry, but with some important differences which undermine its effectiveness.

The most important difference is that the offshore oil and gas industry (fourteen companies) is not subject to the Environmental Management Law (*Wet milieubeheer*), under which all other business activities come. Therefore, offshore oil and gas installations need no environmental permit. Only one regulation is applicable (included in a non-environmental law), which is concerned with banning the discharge of oil-based drilling muds. Exploration and production of oil and gas fall under the jurisdiction of the Ministry of Economic Affairs. Therefore this ministry, with its ideology of non-interference in the market, i.e. no regulation, was the ministry responsible for the agreement. However, the Ministries of Water Management and Environment participated in the negotiation process and have succeeded to a certain extent in improving the design of the agreement. Luckily, the objectives of the agreement were derived directly from the National Environmental Policy Plan and were not themselves negotiated. These objectives are similar to those contained in the chemical industry agreement, i.e. reduction percentages for all emissions relative to 1985. However, and this is another weakness, the trade organisation NOGEPA (Dutch Oil and Gas Exploitation Association) attached critical remarks to many of these environmental objectives. Therefore it is questionable whether industry has a similar perception of environmental problems as have the other parties.

In this particular case, the choice for an agreement meant in practice that the negotiated path was officially preferred instead of regulation, though no arguments have been substantiated why an agreement was considered more effective. This is remarkable, because originally the Ministry of Water Management intended to introduce legislation for this category of industry, though it faced heavy opposition from the Ministry of Economic Affairs. Difficult consultations between the three ministries (the third being the Ministry of Environment) and the sector on necessary environmental measures for the offshore industry had been going on since 1989. In 1992 an Environmental Action Plan was formulated in which a number of important legislative activities were foreseen. But then the idea arose of switching to the negotiated-agreement approach and the parties abandoned the legislative proposals, causing a delay in the implementation of concrete measures of at least three years, but in practice probably in the order of 8–10 years. As third parties had no possibilities for inquiry and appeal because of the lack of a system of environmental permits for offshore platforms, a special procedure has been designed through the efforts of the Ministry of Environment so that third parties may comment on draft CEPs. An appeal procedure is not foreseen, however. Monitoring of the agreement is carried out by the sector itself, instead of by an independent agency, as in the chemical industry (the FO Industrie).

The results based on the first generation of draft CEPs in the offshore oil and gas sector in mid-1996 showed rather disappointing results. While large reductions of sulphur-dioxide (SO_2) and methane (CH_4) emissions have been achieved (meeting the objectives of the agreement), substantial bottlenecks remain to be tackled (in particular NO_x) and water discharges have hardly been dealt with. The successes are mainly achieved through simple technical measures and by the closure of platforms. In spite of the agreement, water discharges are not considered a problem by industry, while NO_x emissions have actually risen due to the need to increase pressure in end-of-life oil and gas reservoirs. If the regulations originally intended had been introduced, the performance of industry would have been much better, because then, for the first time in history, this sector would have been confronted with the same environmental requirements as any other branch of trade.

Other integral agreements

Various flaws have plagued other integral negotiated agreements concluded in the Netherlands as well, though this does not necessarily mean that they do not work. That also depends on the particulars of the situation:

- An agreement with the base-metal industry (1992) meets all of the 'rules of the game' criteria but has been hampered in practice because of the unwilling attitude of most participating companies (i.e. an element of the choice criteria). This has brought significant pressure to fall back on traditional command and control in order to reach the necessary environmental improvement. A further problem is the relatively high costs of substantially reducing emissions, due to the capital-intensive character of this industry. Finally, there have been serious problems with transparency because many companies did not want to give their CEP to third parties. (For details, see Stichting Natuur en Milieu 1994.) In the second round of CEPs the performance of the sector has been greatly improved, though a small number of substantial problems remain (Hagedoorn 1997; FO Industrie 1997).
- Agreements with the printing industry (1993) and the metal and electronic industry (1995) do not meet the element in the choice criteria which states that a limited number of parties must be involved. Both sectors contain too many companies, the metal sector comprising more than 10,000 companies. Though the organisation of the implementation of these agreements differs from those with the chemical and base-metal industry (with packages of measures for particular groups of companies), the results are very difficult to monitor and problems are very difficult to tackle. Finally, the rewards for companies are much less evident then in other agreements. The agreement with the printing industry was scheduled for evaluation in 1997.
- Although the agreement on petrol filling stations (1991) does not meet an element of the choice criteria (number of companies addressed), it works rather well, partly because the agreement has been supported by regulation (at the request of the companies), but also because the sector itself decided to set up a sectoral fund (SUBAT) for soil sanitation, the most pressing environmental problem of the sector (besides emissions of volatile organic compounds (VOC)). This fund is generated from a small rise in the price of petrol. Filling stations which decided to close down before a certain deadline (often the ones owned by a private person who could not afford soil sanitation) could apply to the fund, which then would take on the responsibility of paying for the soil sanitation. Very important in this case has been the role of the major oil companies, which own a large portion of the petrol filling stations and whose role

will increase through this restructuring of the sector. (More information is found in Stichting Natuur en Milieu 1993; Ingram, forthcoming)

Experiences with single-issue agreements

Most of the many negotiated agreements in the member states of the European Union are concerned with *single environmental issues* (for a recent overview of existing European agreements, see Öko Institut *et al.* 1996; European Commission 1997). One can think of particular waste streams like batteries, old tyres, wrecked cars, packaging, emissions of volatile organic compounds, chlorofluorocarbons (CFCs), energy efficiency (or CO_2 emissions), soil sanitation, tropical timber, and so on. The important thing to realise with this type of agreement is that, in contrast to the integral agreements described above, they are usually not meant to be a communicative or management instrument targeted at an integral improvement of companies' environmental performance. Rather, they are usually meant as an instrument for reaching a specific environmental goal where regulation is very complicated or even hardly possible, or (usually in the bad cases) as an alternative to regulation. A number of modern Dutch cases illustrate the limitations of poorly considered single-issue agreements as well as the potential environmental gains from agreements which adhere to the essential rules of the game.

Energy efficiency

A particularly striking example is the large group of single-issue agreements on energy efficiency which have been concluded since 1992. The responsible ministry is the Ministry of Economic Affairs, which has no high environmental ambitions and prefers a no-regrets policy in this area. This is the main reason for their choice of the instrument in the first place. There are more than thirty of these agreements, but some eighteen for industrial sectors. The energy-efficiency agreements violate several elements of the aforementioned rules. Although the choice of the instrument might be justified in this particular case, the problems lie in the ambitions and the design criteria.

The ambitions of the agreements are low, despite enormous scope for energy savings in industry. The present objectives barely exceed the average increase in energy efficiency which industry already achieves by itself (1.0–1.2 per cent per year compared to the 1.6–1.8 per cent per year required by the agreements). As a rule, measures considered under the agreements must have a maximum payback time of three to five years, which actually means normal profitable business practice and nothing more. This is also what is meant by a 'no-regrets' policy. Studies show that the economic potential (cost-neutral options) and technical potential (all known options, without reference to costs) are each much higher (World Wide Fund for Nature 1996): efficiency gains of between 24 and 29 per cent for economic potential and 37 per cent for technical potential are possible between 1990 and 2000. These percentages increase substantially if the time horizon is extended.

Moreover, efficiency targets do not limit the CO_2 emissions in an absolute sense. Although the official intention of the government is to reduce CO_2 emissions by 3 per cent by 2000 compared to 1989 levels, due to the central position of the Ministry of Economic Affairs in energy policy, energy-efficiency objectives were chosen instead of CO_2 reduction targets. Between 1992 and 1996 the government lowered the national industry targets for 2000 (relative to 1989) from a 20 per cent to 16 per cent efficiency increase, because of lower economic growth. However, CO_2 emissions for 2000 are predicted to rise by 7 per cent, instead of dropping by 3 per cent, due precisely to expected economic growth (RIVM 1996).

In an evaluation done by Stichting Natuur en Milieu (1996), the noncommittal form of the energy-efficiency agreements is an aspect that is increasingly problematic for sectors of industry lagging behind in the schedule. For example, the paper and cardboard industry, the rubber industry and many branches within the food industry are considerably behind schedule (Ministerie van Economische Zaken 1996). Only when considerable efforts are made will these sectors be able to reach their objectives for 2000. This is not a problem of technical possibilities, as has been mentioned, but mainly a lack of attention paid to the problem within the companies. There is no real stick behind the door in order to deal with free riders. Although since 1992 authorities may prescribe energy demands in environmental permits, companies participating in one of the energy agreements are exempted from this permit obligation, provided that their efforts are in line with the objectives of the agreement. This is also the reward for companies participating in the agreements.

Also, permit-granting authorities first have to check whether a company adequately fulfils its agreement obligations. However, the lack of transparency goes very far, because not even the authority gets access to the actual data of the company; it has to rely on the opinion of an intermediary organisation, NOVEM (Netherlands Agency for Energy and the Environment), which carries out all monitoring of the energy agreements. If NOVEM endorses the performance of the company, the authority cannot prescribe energy measures in the permit. NOVEM, for its part, relies for data on the companies themselves, in some cases even on the trade organisation which collects the data, and there is no verification. There are known cases where a company performed badly but received the consent of NOVEM (for example Triton Paper). In those sectors which fall behind schedule, one would expect that companies would have measures prescribed in their permit, but there are no indications that this really happens. It is possible that this will occur only if the sector does not improve its performance in the period 1995–2000.

The lack of transparency naturally extends also to the public. Though companies have to produce a company energy plan under the articles of the agreement, even an abstract of this plan becomes public only if there is a new environmental permit application. Several NGOs tried to get access to the information in the hands of NOVEM through the court by asking it to apply the law on open government practices. In December 1994 the Court held that, formally, the Ministry of Economic Affairs could not supply the data, because NOVEM held them. NOVEM did not want to provide them because it was not a partner in the agreements. The court went on to say that this construction was culpable and contrary to the meaning of the law on open government practices. However, in practice nothing could be done and nothing has changed since then. The lack of access to the NOVEM data has also been criticised by researchers wanting to control its work. Most of the critical comments have been substantiated in an extensive evaluation report published by the Ministry of Economic Affairs (Rijksuniversiteit Utrecht 1997).

In the EU several other weak negotiated or voluntary schemes for industrial energy efficiency have been signed. In the United Kingdom, for example, there is a highly ineffective Making a Corporate Commitment Campaign, which only asks industry to sign a declaration which contains seven principles (see Chapter 2). These seven principles do not contain quantitative objectives and it is not obligatory for the participating companies to implement them. In surveys of the Department of the Environment there are strong indications of the ineffectiveness of the scheme. Still, the UK government considers the declaration one of its major policy elements in reaching CO_2 reduction targets (Jenkins 1995). Similarly, German industry designed a number of *Selbstverpflichtungen* in 1996 for the reduction of CO_2 (see Chapter 3). According to experts within the Dutch chemical industry, the way these *Selbstverpflichtungen* are monitored gives a meaningless picture of the actual increase in energy efficiency.

Packaging

Another example of a single-issue agreement is the packaging agreement signed in 1991 between the Ministry of Environment and the Stichting Verpakking en Milieu (Foundation for Packaging and Environment), representing approximately 150 companies. The targets of the agreement, specified for five types of packaging waste (glass, ferrous metals, non-ferrous metals, paper and plastics) were not very ambitious: in particular, the general target of only a 10 per cent reduction in packaging waste over ten years was rather low. The agreement also established an overall minimum recycling target of 60 per cent in ten years (50–80 per cent depending on material type) and included interim (halfway) objectives for 1995. Still, the agreement led to action in an area where effective regulation of re-use and recycling was not possible, because general legislation for a very diverse range of packaging products is very difficult to formulate and control. The targets of the agreement have been negotiated by industry to a certain extent: the reduction percentage is now related to what is achievable with current industrial improvement trends, rather than a more ambitious reduction goal.

Initially, monitoring of industry's performance was chaotic and it was very difficult for third parties to obtain data. Several years after the signing of the agreement this situation has significantly improved, not least through the influence of environmental organisations and the strong interest of the press. Now, both industry and an independent research organisation (RIVM) monitor annual progress and publish the results. Information about individual company performance, however, is still lacking.

The weakest point of the agreement is the unpunished free-rider behaviour of a group of companies, which, particularly for plastic waste, is the main reason why several of the targets for specific waste streams have not been achieved (although other targets have, particularly those for waste streams where effective collection systems have been in place for many years, i.e. paper and glass) (Commissie Verpakkingen 1995). No provisions have been made in the agreement to tackle this problem.

The EU Directive on Packaging Waste, which came into force in 1994, might solve this problem to a certain extent. This Directive contains very unambitious targets, at least much less ambitious than the Dutch packaging agreement (the EU law requires at least 15 per cent recycling for every stream of packaging waste).[4] The point is that the Directive legally requires a government to address producers of packaging material. In October 1996 the EU packaging Directive was implemented in Dutch law and, consequently, the existing packaging agreement was superseded. Negotiations were still going on with industry, with the goal of continuing the agreement in another form. The problematic point here is that the Dutch government is hampered in its negotiating position because it can no longer threaten to adopt recycling measures which go much further than the EC Directive.[5] The intention is now to exempt companies from the law if they join a new packaging agreement. This gives them some flexibility as they are not bound by strict rules, but as a consequence they have to perform better than is required under EC law. If they do not perform better, the law is there to guarantee minimum performance, and measures for individual companies can be enforced.

Similar or (more often) worse experiences with packaging agreements have been noted in several EU member states, with the UK as a documented disaster. The extremely sluggish reaction of British industry to repeated government calls to present a packaging plan meeting five key criteria indicated the problems to come. At a certain stage industry itself even asked the government to come up with regulations, in order to overcome internal problems (Eden 1997). In the period 1990–5 not even a concrete plan of action was decided upon, by government or by industry (Jenkins 1995; Eden 1997).

Other single-issue agreements

Flaws have plagued many single-issue agreements in other areas of Dutch environmental policy:

- An agreement on plastic-waste reduction in industry (1993) does not comply with elements of the ambition, choice and design criteria listed above: after three years of negotiation this agreement said only that in 1995 industry should have finished research into the possibilities of prevention and re-use of plastic waste within companies. Objectives are lacking, despite the fact that in 1988 the Ministry of Environment had already formulated a concrete target. After several years of implementation, this agreement is still hardly known in industry.
- An agreement on the reduction of pesticide use is far behind schedule because of severe failures with regard to elements of the choice and design criteria. The lack of adequate regulation (in particular the admittance policy of existing and new pesticides) and/or other substantial incentives like taxing pesticides is the main cause of the failure of this agreement. Many more details are provided in a joint report by a large group of national and regional NGOs (Muilerman and Steekelenburg 1996).
- An agreement between the national water authority and Hydro Agri (a producer of artificial fertilisers) (1987) did not meet elements of the choice and design criteria. As a consequence, in the permit procedure existing legal requirements were violated. Therefore the agreement was nullified in a court procedure prompted by two NGOs (see Biekart 1995 for details).
- A really effective agreement was signed on the reduction of acidifying emissions from electricity plants (1990). This agreement meets the criteria and it works. The main point is that the electricity companies themselves may decide whether a site is subjected to further emission control, as long as the targets for the sector as a whole are met. This is purely a matter of cost-effectiveness. The agreement is supported by legislation on emission standards for electric power plants.

Reports for the EC on negotiated agreements (Öko Institut *et al.* 1996) and other information sources illustrate the meagre results achieved by single-issue agreements in several member states, as well as in other countries. Agreements which work are almost inevitably supported on one or more points by legislation. In the other cases, agreements usually do not work well, or work only because specific pressures have been effective:

- A good example of the necessity of regulatory support is Denmark, with its agreement on the recycling of car batteries. Here, after an agreement had been signed, an extensive set of rules and regulations developed in the course of time in order to cope with a range of loopholes which allowed free riders to circumvent the agreement. With these in place the agreement seems to work (Pedersen and Elmvang 1996).
- An example of effective external pressures is the agreement in Germany on the reduction of the use of CFCs in refrigerators, earlier than legally required. Through the introduction of the CFC-free refrigerator ('Greenfreeze') by a joint initiative of Greenpeace and DKK Scharfenstein, supported by the legislation of the environment minister, Töpfer, all leading producers – such as Bosch-Siemens and Liebherr – had CFC-free refrigerators on the market within six months of claiming that commercialisation of such a technology was not possible.

Our examples illustrate that negotiated agreements, integral or single-issue, must obey the rules of the game, or run the serious risk of failure. Only when the authorities (or government) involved play their negotiating role adequately may negotiated agreements become a useful instrument, creating commitment and cooperation between industry and government parties.

The current experiences with negotiated agreements, integral or single-issue, can hardly prove that industry performs better with than without them. Often the contrary seems to be true. In the shorter term the benefit is in the process, not in the performance: it calls for industry to take its responsibilities and it might integrate environmental thinking in business practices. In the longer term that might change and real results might be achieved, but only if the process works well enough. As some of the previous examples illustrate, agreements designed according to the rules of the game (in particular the integrated ones) might provide fertile ground for more strategic environmental thinking in industry, on the condition that government also takes the necessary action.

Pros and cons: frequently heard arguments

This section devotes its attention to a number of important issues with regard to negotiated agreements which could not be dealt with in earlier sections. Many of these issues are mentioned in the international discussion on negotiated agreements.

In the view of some people, *command and control-type regulation has proved to be ineffective* and therefore new instruments like negotiated agreements are necessary to create a more effective environmental policy. Though there is some truth in this view, it is far too simplistic. While a great deal of environmental legislation is not effective, this has nothing to do with the instrument itself, but finds its root causes sometimes in the too detailed and complicated design of legislation, and more especially in the lack of money and political attention devoted to implementation and enforcement.

The implementation and enforcement of negotiated agreements *require capable, flexible and creative people* who can negotiate and distinguish between features and details. They require the permanent training of people. These are high standards, but the point is that negotiated agreements do not require less. On the contrary. We have seen in the Netherlands that the introduction of negotiated agreements requires a change in the culture of permit-granting authorities and enforcement personnel that will go on for many years to come. Similarly, the same switch must be made by companies which could previously wait to see what was being asked of them and react. Now they have to become proactive and make their own plans – a real shock for many of them. This is very useful indeed, but it is not easier than a command and control approach.

In some EU member states (like the United Kingdom), and in Brussels itself, an important argument in favour of negotiated agreements is that they *entail lower costs* – lower costs for regulation and lower costs for industry. This can be an argument only of a secondary nature. To begin with one must look at environmental effectiveness, otherwise one misses the point. This view is supported by the simple fact that the environmental effectiveness of agreements is usually monitored but their economic efficiency is not. However, if one looks at the successful agreements in the Netherlands one may conclude that, after an initial investment in people and training to carry out negotiated agreements, cost benefits may be found in the procedure of granting permits, in particular for companies. At least two chemical companies have tried to estimate cost benefits for themselves (Hoechst in Vlissingen and DSM in Geleen). They found results in the order of 10 per cent savings in time (= costs) spent. The reason for these savings lies in dealing with all the environmental problems of a site in an integral manner. The cost-savings argument is related to efficiency and naturally does not hold for the level of the environmental investments of a company. This is supported by a cost-effectiveness analysis of eight international cases (Öko Institut *et al.*, forthcoming).

For permit-granting authorities, no quantitative data are available on the efficiency of agreements in terms of manpower. As was observed previously, agreements seem to increase the need for qualified personnel. An advantage of the agreements might be that it becomes much clearer which companies can bear environmental responsibilities, while others show that they need a straightforward approach and strong enforcement. At least one province in the Netherlands uses a rating system for the quality of companies' environmental practices, which has consequences for its enforcement priorities.

With regard to the *undemocratic nature* of negotiated agreements, the argument holds in principle, but the problem can be solved by following the 'rules of the game'. Agreements which are not designed according to these rules are very often undemocratic indeed. Generally, the reason for not following these rules is that the parties involved wish to make it impossible for any third party to track the results of the agreement, whether the third party is an environmental NGO, a citizen or parliament.

In the author's experience, negotiated agreements may become *an obstacle to ongoing environmental policy and the creation of new instruments*. The agreement with the oil and gas industry which blocked the introduction of necessary legislation has already been mentioned. The way the ecotax on energy is being treated is also very illuminating. Industry as a whole is strongly opposed to this idea, for obvious reasons. One of its arguments is that it has made a deal for the improvement of energy efficiency. In the opinion of industry, an energy tax thwarts its voluntary efforts instead of supporting them. Trade organisations have therefore introduced a clause in the energy agreements that the introduction in the Netherlands of an ecotax on energy can lead to the termination of the agreement. In general, Dutch employers' organisations argue continuously that government should renounce the introduction of new environmental policies, legislation and instruments because agreements containing environmental objectives have already been signed, thus making any new instrument superfluous (see Chapter 4). This can be seen, in particular, from the discussion around the groundwater tax introduced in 1994 (de Graaff 1993).

The question is often put whether agreements *block or invite industry to innovate*. Again, the answer will depend on the particular case being considered. In general, agreements which meet the rules of the game will be able to stimulate innovation, provided that other instruments are in place to overcome particular difficulties – particularly financial stimuli, a strong research and development sector, within or outside industry, and an effective network of intermediate organisations which is working directly with researchers and companies. If this infrastructure is lacking, companies, especially small and medium-sized enterprises (SMEs), will not tend to innovate. In the complicated process of pushing companies to innovation, negotiated agreements have only a modest role.

Conclusions

This final section evaluates the four propositions put forward in the introduction to this chapter and then offers some final remarks regarding the use of negotiated agreements in the European Union.

With regard to the first proposition, no example is known to us of an agreement that represents a major step towards sustainability. The ambitions of the agreements we have seen are usually compliance with legislation and/or somewhat more ambitious targets in the longer term. That is to be expected, because the actual balance of power between government and industry will not allow more ambitious objectives than would be achievable through other means where all parties are consulted, i.e. legislation. Agreements, therefore, will not solve the major dilemmas in environmental policy, though when they are used in the best possible way they do contribute towards sustainability.

With regard to the second proposition, although there are examples of agreements initially designed without a regulatory framework, in many of these cases this framework is constructed in a later phase (often at the request of companies themselves) in order to cope with free riders. In the cases where this has not been done there have been real problems with free riders.

With regard to the third proposition, access to information provides a strong impetus for companies not to stay behind, but rather to stay even with the pack, or even perform better than others for reasons of market opportunities or public relations. This is a major reason for some environmental NGOs (like that of the author) to support negotiated agreements in particular cases and it is a unique possibility for government to organise pressure on companies to perform better or to refrain from free-riding.

With regard to the fourth proposition, lack of credibility damages the value of an agreement in the eyes of the public, politics and the press. Especially crucial is the credibility of the position the governmental party takes in negotiating, designing, implementing and enforcing an agreement. There are no examples known to the author of non-credible agreements that have had good results. A detailed look at what first seemed like examples to the contrary revealed cases where credibility had improved over one or two years. We feel supported in these conclusions by the findings of international research (European Environment Agency 1997; Öko Institut; Stichting Natuur en Milieu and FIELD, forthcoming).

Some final words about the application of negotiated agreements in the European Union are apposite here. It is useful to mention that discussion focuses on three types of agreements: those at the Community level, those in member states as a possible means to implement certain types of EC Directives, and purely national agreements. The thrust of this chapter has been that agreements in general have no value in improving the effectiveness of environmental policy in Europe unless the policy context in Europe is much improved.

Why such a pessimistic view? First, the possibilities of negotiating agreements which follow the 'rules of the game' with regard to ambition, choice and design are rather few indeed. Frequently agreements will have no additional value, as they demonstrably should have. Second, the implementation and enforcement of many agreements will probably lead to a number of new problems, like distinguishing between companies which are performing well or badly and defining adequate enforcement. Also, if agreements meet our 'rules of the game' a major shift in attitude and qualifications of both civil servants and industry personnel becomes necessary in order to reach good results. Third, agreements will only work in optimal form if they are used in addition to other new instruments which fit into the philosophy of putting greater responsibilities on the shoulders of industry. Some possibilities were suggested above, but there is no reason to trust that such instruments will become available in the near future.

Still, it might be worthwhile experimenting a little with negotiated agreements, both at the Community and at the member-state level. The main problem is not a lack of belief in the potential value of negotiated agreements which meet the rules of the game. Rather, one wonders how governments will apply these rules and whether they will take a firm position when problems arise. In this sense, negotiated agreements do not seem to differ very much from environmental regulations. The Communication of the European Commission to the Council and the European Parliament in Environmental Agreements (COM (96) 051 Final), issued in November 1996 and containing reference to a large number of our rules of the game, is therefore quite important.

Notes

1 With regard to negotiated agreements, the Netherlands Society for Nature and Environment critically follows their development and implementation and signals parliament, government, public bodies, industry and the press when important issues arise. The society does not participate in any of the agreements.

2 Information can be found regularly in the newspaper of the Working Group on Community Right-to-Know in Washington, DC.

3 Many studies have tried to identify the actual policy changes required to achieve sustainable development. What is actually necessary is well described in Reijnders (1996).

4 For a discussion of the Directive's development and its relatively weak requirements compared to various pre-existing national recycling schemes, see Golub (1997).

5 In this case the relative *laxity* of EU standards constrains the use of voluntary agreements; although the Directive allows member states to establish more stringent recycling targets, a number of conditions must be satisfied (e.g. self-sufficiency, proximity) and there is a risk of violating EU rules on free trade. In other cases, however, it is the possibility of *stringent* EU rules which acts as a constraint by injecting considerable uncertainty during the negotiating stage of national voluntary agreements (see Chapters 1 and 5).

References

Algemene Rekenkamer (1995) *Convenanten van het Rijk met Bedrijven en Instellingen* (The Hague: Algemene Rekenkamer).

Barracha, F. (1996) 'Voluntary compliance as a means of adapting to environmental legislation', contribution of the Portuguese minister of environment in Lisbon to the EC Workshop on Voluntary Agreements, Brussels, May.

Biekart, J. W. (1995) 'Environmental covenants between government and industry: a Dutch NGO's experience', *Review of European Community and International Environmental Law* 4(2): 141–9.

Bogie, M. J. S. (1993) 'Het convenant als sturingsinstrument in het milieubeleid', *Openbaar Bestuur* 11: 28.

Commissie voor de Toetsing van Wetgevingsvraagstukken (1992) *Convenanten* (formal advice for the Dutch government on covenants) (The Hague).

Commissie Verpakkingen (1995) *Jaarverslag* (Utrecht).

de Graaff, J. J. (1993) 'De terugtredende overheid?' *Milieustrategie* 1: 14.

Dietz, F., Vollebergh, H. and de Vries, J. (1995) *Environment, Incentives and the Common Market* (Dordrecht: Kluwer Academic Publishers).

Eden, S. (1997) 'The politics of packaging in the UK: business, government and self-regulation in environmental policy', *Environmental Politics* 5(4): 632–53.

Europe Environment (1995) 4 April, no. 452.

European Commission (1997) *Study on Voluntary Agreements Concluded Between Industry and Public Authorities in the Field of the Environment*, unofficial report (Brussels: DG III).

European Environment Agency (1997) *Environmental Agreements: Environmental Effectiveness*, vols 1 and 2 (Copenhagen: EEA).

FO Industrie (1997) *BMP-2 Rapportage Basismetaalindustrie* (The Hague: FOI).

Golub, J. (1997) 'State power and institutional influence in European integration: lessons from the Packaging Waste Directive', *Journal of Common Market Studies* 34(3): 313–39.

Greenpeace International (1992) *The Greenpeace Book of Greenwash* (Amsterdam: Greenpeace).

—— (1995a) *Strategies to Promote Clean Production, No. 4: Extended Producer Responsibility* (Amsterdam: Greenpeace).

—— (1995b) *Strategies to Promote Clean Production, No. 3: Ecological Tax Reform* (Amsterdam: Greenpeace).

—— (1996) *Voluntary Agreements* (Brussels: Greenpeace European Unit).

Hagedoorn, N. (1997) 'Verzuring blijft heet hangijzer in basismetaalindustrie', Natuur & Milieu 21 (7/8): 8–10.

Ingram, V. (forthcoming) 'Netherlands, case study 1: the Subat Agreement', in Öko Institut, Stichting Natuur en Milieu and FIELD (eds) *New Instruments for Sustainability: The Contribution of Voluntary Agreements to Environmental Policy* (Darmstadt: Öko Institut).

Inspectie Milieuhygiëne (1995) *Landelijk Produkt Evaluatie BMP's Chemische Industrie: Onderdeel Tussenrapportage over het Opstellen van Bedrijfsmilieuplannen* (The Hague).

—— (1997) *Evaluatie van de Bedrijfsmilieuplannen in de Chemische Industrie* (The Hague: Inspectie Milieuhygiëne).

Inter Provinciaal Overleg (1996) *Evaluatie Doelgroepmanagement Industrie: Basismetaal, Chemie en Zuivel. Op weg van de 'Eerste Generatie' naar de 'Tweede Generatie' Bedrijfsmilieuplannen* (The Hague).

Jenkins, T. (1995) *A Superficial Attraction: The Voluntary Approach and Sustainable Development* (London: Friends of the Earth).

Ministerie van Algemene Zaken (1995) *Guidelines for Covenants (Regulation No. 95M009543)* (The Hague: Ministerie van Algemene Zaken).

Ministerie van Economische Zaken (1996) *Meerjarenafspraken over Energie-efficiency. Resultaten 1994* (The Hague).

Ministerie van Verkeer en Waterstaat (1995) *Doelgroepenbeleid: Ontlasting voor Water?* (The Hague: Directoraat Generaal Rijkswaterstaat).

Ministerie van Volkshuisvesting, Ruimteliijke Ordening en Milieubeheer (1994) *Provisional Code of Conduct for Concluding Environmental Covenants* (The Hague).

Muilerman, H. and Steekelenburg, A. (1996) *De Buik vol van Gif. Tussentijdse Evaluatie van het Meerjarenplan Gewasbescherming (MJP-G) Door Samenwerkende Milieu-organisaties* (Rotterdam: Zuid-Hollandse Milieufederatie).

NCI (1995) *Nederlandse Chemische Industrie*, March.

OECD (1996) *Council Recommendation on Implementing Pollutant Release and Transfer Registers* (Paris: OECD).

Öko Institut, Stichting Natuur en Milieu and FIELD (Foundation for International Environmental Law) (1996) *Draft Report on the Experiences with Voluntary Agreements in Seven EU-member States, Poland and the United States* (Darmstadt: Öko Institut).

—— (forthcoming) *New Instruments for Sustainability: The Contribution of Voluntary Agreements to Environmental Policy* (Darmstadt: Öko Institut).

Pedersen, K. B. and Elmvang, M. (1996) 'Voluntary and negotiated agreements on lead accumulators in Denmark', contribution of the Danish Environmental Protection Agency (EPA) to the workshop of the European Commission on Voluntary Agreements, Brussels, May.

Reijnders, L. (1996) *Environmentally Improved Production Processes and Products: An Introduction*, Environment and Management Series, vol. 6 (Dordrecht: Kluwer Academic Publishers).

Rijksuniversiteit Utrecht (1997) *Afspraken Werken: Evaluatie Meerjarenafspraken over Energiebesparing* (Utrecht: Rijksuniversiteit).

RIVM (National Institute of Public Health and the Environment) (1988) *Concern for Tomorrow* (Alphen aan de Rijn: Samson H. D. Tjeenk Willink).

—— (1996) *Milieubalans 1996. Het Nederlandse Milieu Verklaard* (Alphen aan de Rijn: Samson H. D. Tjeenk Willink).

Stichting Natuur en Milieu (1993) *Bedrijfstak Tankstations: Bijlage 8 bij het Rapport 'De Uitvoering van het NMP en de Industrie'* (Utrecht).

—— (1994) *De Basismetaalindustrie en het Doelgroepenbeleid Industrie: Analyse van Proces en Resultaten op weg naar 2000* (Utrecht).

—— (1996) *Multi-year Agreements on Energy Efficiency: A Matter of Trust* (Utrecht), summary report.

UNEP (1994) *Company Environmental Reporting. A Measure of the Progress of Business & Industry Towards Sustainable Development*, Technical Report no. 24 (Paris: United Nations Environment Programme/Industry and Environment).

WICE (1994) 'Improving policy cooperation between governments and industry', report of the Working Group on Policy Partnerships (Paris: World Industry Council on Environment).

Winsemius, P. (1993) 'Environmental covenants and contracts: new instrument for a realistic environmental policy', *Tijdschrift voor Milieu Aansprakelijkheid* 4; 89.

World Wide Fund for Nature (1996) *Policies and Measures to Reduce CO_2 Emissions by Efficiency and Renewables: A Preliminary Survey for the Period to 2005* (Zeist, the Netherlands: WWF Climate Change Campaign).

9

ECOLABELS IN EU ENVIRONMENTAL POLICY

Eva Eiderström

Introduction

Consumers and their strength as market actors have become an increasingly interesting alternative to command and control measures when dealing with environmental problems. As a consequence of the withdrawal of legislation as a steering mechanism, other market-based instruments like voluntary agreements, standardisation, tradeable permits and ecolabelling have become the politically correct method to employ. When these instruments are applied properly, under the correct conditions, they will deliver environmental returns at a much higher pace than traditional tools, but as of yet there are still problems with the way they have been developed and used.[1]

This chapter begins by describing what ecolabelling is from a theoretical point of view. The subsequent section then discusses the EU label, 'the Flower', in more detail. This is followed by three case studies of how a private ecolabel scheme has been used in Sweden. In light of these cases, the final section concentrates on some of the prerequisites which are necessary in order to operate a successful scheme.

The chapter draws upon seven years of experience with a Swedish private ecolabelling scheme called 'the Good Green Buy'.[2] The author has also participated as a representative of environmental organisations in the Ecolabel Forum, the interest groups' consultative forum within the EU ecolabelling scheme.[3]

The current increasing interest in market-based steering instruments like ecolabelling can only be seen in light of the failure of legislators or politicians to combine deregulation and environmental improvements. In Sweden, for example, the so-called ecocycle legislation was introduced in 1994. It included 'extended producer responsibility' which built on voluntary agreements as a new policy tool, compared to prescriptive legally defined enforcement of legislative goals.[4] The political focus on free trade and fewer obstacles to trade necessitates solutions where consensus models are utilised.

This is the rationale behind the 'new approach',[5] in which framework legislation and standardisation are combined to reach harmonisation goals.[6]

Official ecolabelling schemes are usually modelled on standardisation, which results in meagre successes during the first few years. Unclear objectives and unclear environmental priorities have also added to the low cost-efficiency of most official schemes, due in part to the novelty of the instrument and the uncertain prospects of a market-based instrument. It is also understandable that if an instrument is modelled on standardisation and its handling is left to standardisation institutions the knowledge of how markets work is something which will have to be developed over the course of several years.

What is ecolabelling?

According to current International Standards Organisation (ISO) definitions, ecolabelling measures can be grouped in three categories.[7] Type 1 is described as *third-party practitioner schemes*. Establishment of criteria as well as their subsequent evaluation is undertaken by a third party not directly commercially dependent upon the outcome of an applicant's evaluation. The process leading up to establishment of criteria has to be transparent and principles for this are elaborated within the ISO system. This type of ecolabelling is the one most commonly discussed and is consequently also the kind of labelling concentrated upon in this chapter.

Type 2 labelling is described as *self-declaration*, meaning a situation where a producer uses a phrase combined with some logotype to describe some environmental quality in its product without having this evaluated by a third party. Standardisation aims to minimise the number of statements and symbols used by devising a reduced number of globally defined symbols. Examples of statements currently flourishing are 'biodegradable' and 'recyclable'.

Type 3 labelling is also a third-party evaluation scheme – but without reference to established criteria based on relative environmental performance among products in a sector. This type of labelling was developed in California by Scientific Certification Systems. The system relies on making a life-cycle assessment (LCA) study of the applicant's product. A reduced number of the resulting parameters are then depicted graphically on the product in a standardised manner. This is comparable to giving an environmental fingerprint of the product.

To clarify, it should be stressed that the current proliferation of 'green claims' in the shape of pictograms or environmentally formulated phrases on products has nothing to do with the above-mentioned types of ecolabelling. Most green claims are not based on harmonised or standardised parameters but highlight what the producers consider the important environmental quality of their product. Each of the three types of ecolabels listed above aims to reduce the proliferation of green claims in order to give consumers a more objective basis for decision-making in terms of environmental quality.

Type 1 labelling: how are most schemes organised?

The best-known ecolabelling programmes belong to the group of 'official schemes', connected with national governments in terms of financing, control or ministerial governance. Mostly the actual practitioners are the national standardisation bodies, which have specific groups within their organisation that deal with the establishment of criteria, the evaluation of applications and the granting of licences for the label. Other varieties include ministerial officials designated to function as ecolabelling officials, or independent boards operating secretariats. Most official schemes depend heavily on government funding. Although most schemes are designated as self-financing, the granting of licences and income in terms of licensing fees remains very meagre, at least during the initial years. Even in the case of the relatively successful Nordic ecolabelling scheme, 'the White Swan', after six years of operation the licence fees account for only two-thirds of the annual budget. In practice, one can conclude that ecolabelling will never become an instrument whose cost is totally recouped by the licence fees on labelled products.

The ecolabelling scheme which the author's organisation (the Swedish Society for Nature Conservation – SSNC) operates is of the private third-party practitioner type, but has no connection to the government in terms of financing or governance. In this case the scheme is financed by SSNC and three retailers operating in the Swedish market. The SSNC has a mandate to establish criteria independently, but up to the final decision relies on the same kind of open and transparent process found in any official scheme. It evaluates applications and grants licences to use the logotype, a peregrine falcon and the phrase 'Good Green Buy – this product is in accordance with criteria established by the Swedish Society for Nature Conservation'.

What is the normal procedure for operating a general ecolabelling scheme?

Operating an ecolabelling scheme involves defining criteria for the label and then licensing it. The basis of labelling products in the market is evaluating them on a number of environmental parameters for a specific product group. These parameters measure the environmental impact caused by a functional unit of the product in question. How this functional unit is defined varies between schemes, but normally the production stages are largely ignored and emphasis is placed on the product's inherent environmental effects – e.g. the amount of recycled paper content in a brand of toilet paper or the amount of solvent in a solvent-based paint.[8] Usually the aim of the label is to diminish the impacts caused by the products during their use and disposal; less often, it is to reduce the impacts caused by their production. This is a consequence of the political difficulties of promoting specific production techniques across borders. The questions of technical barriers to trade and protectionism disguised as environmental demands have received considerable attention within international organisations such as the Organisation for Economic Cooperation and Development (OECD), the United Nations Environment Programme (UNEP) and the United Nations Conference on Trade and Development (UNCTAD 1995; OECD 1995; European Commission 1996c; World Trade Organisation 1995: chs 1 and 11; Golub 1998; Vogel 1998).

Mostly, criteria have a tendency to be based on a matrix, whereby environmental demands are calculated for groups of weighted parameters and the sum of these scores then constitutes the actual environmental 'cost' of the product. Sometimes this matrix is accompanied by hurdles, minimum standards which diminish the possibility of fully compensating for high environmental impacts in one area by concentrating on reducing the impact of another parameter.

Most schemes endeavour to take into consideration all environmental impacts caused during the entire life-cycle of a product.[9] In the early days of most schemes this was also based on hopes that LCA would provide an objective basis for the definition of criteria. All LCA models have the difficult task of balancing incompatible entities, which is usually solved by attaching weights to the parameters measured, resulting in indices or points. These weights differ substantially according to who performed the LCA, who ordered it and in what country it was made.[10] However, there is still no consensus as to what should be taken into account in the LCA models currently in use.[11] The pragmatic solution to this has become the use of Life-cycle Inventory (LCI) as a tool to find parameters relevant to each product group.

The definition of criteria has to be an open and transparent process to which all stakeholders have access. 'Access' means the right to get material and the right to register one's opinion. Access does not guarantee influence on the final outcome, however, since most ecolabelling schemes are modelled on standardisation, where consensus decision-making is the norm. In reality this means that whoever has the resources to devote time and effort to the criteria-definition process can do so. The system itself does not make provisions to ensure a balance among parties controlling different levels of resources.[12] Most schemes, therefore, are very heavy on industry input and very light on input coming from environmentalists or consumers.

Criteria are usually defined in several steps. The first step is when a draft is presented stating the main impacts that can be dealt with, and also suggesting hurdles and values for parameters. This draft is then circulated and debated in expert groups and among interested stakeholders.

Suggestions and proposals are collected and a final version is prepared. In practice this can be a process that takes years to accomplish. Usually there is more than one round of hearings and perhaps several hierarchical levels which exercise influence. There is usually a high degree of industry intervention in the form of meetings and seminars where industry views are presented to the group working with the criteria or the board in charge of making the final decision.

In a second stage producers, retailers, importers and others can apply for a licence to use the label. Foreign producers apply to the country devising the label. The applications are usually based on performance tests on the specified environmental parameters and product performance tests. Evaluations are made by the practitioner organisation and there is usually quite a detailed procedure for assessment to ensure that no product is granted a label incorrectly.[13] This also means that the evaluation can take a considerable time to perform and usually implies that producers have to incur a high cost in order to be able to submit a complete application, a factor which might significantly deter use of this instrument. It is hard to find documentation on how long the average application takes since the evaluation of products is not an open process. There are many reasons for this. In some instances applying for a label represents a change in marketing strategy which the applicant wants to conceal from competitors. For most, though, the uncertainty over how the application will be rated is the main reason for being secretive.

If evaluated and accepted, the product gets a licence to use the label of the ecolabelling scheme. The producer will enjoy improved marketing through the positive image associated with the logotype itself, as well as from a sign that states that the product, in comparison with other similar but unlabelled products, represents an environmentally preferable choice. The label does not guarantee superior performance or quality in a more traditional sense. Nor does it give any indication on the pricing of the product.[14]

The EU Flower

The EU ecolabel scheme operates under Regulation 880/92, which contains procedural guidelines for establishing criteria. The Commission itself has a high degree of influence over criteria formulation since it can devise them independently of the competent bodies of the member states (which are themselves called for by the Regulation) and independently of the Council, unless the latter, within three months, acts by qualified-majority vote (QMV). The competent bodies are mandated by the Commission to be responsible for the handling of the scheme in the member states. Usually the competent bodies are also responsible for the national labels in the countries where they exist.

Prior to 1996 the Commission mandated a member state to be the lead country in developing draft criteria for a certain group of products. The competent body of the country then appointed a working group in which representatives from stakeholder groups and experts from other member states could participate. The expert group developed a draft, which was then presented to the Commission. On the basis of this draft and discussions in competent-body meetings (preceded by a meeting of the Ecolabel Forum where interest groups formulate their opinion on the draft),[15] the Commission made a proposal for a criterion. This final version was then voted on by the member states in a Regulatory Committee. If a majority of the members in the committee voted for the Commission proposal, the Council, on proposal from the Commission, adopted the criterion. In the case of a blocking minority the Commission had two options, to reformulate and succeed in getting majority support or to take the proposal straight to the Council.

Since 1996, however, this procedure has been slightly modified.[16] The Commission itself has taken a much more active role in formulating the draft criteria by delegating the work directly to consultancy firms. Any interested party can participate in the working groups attached to the product groups and there is no longer a formal lead country. Most of these 'new' product groups have produced a draft, so it seems that this change of procedure will yield criteria more quickly than the earlier model.

Regulation 880/92 states that a review shall take place within five years of its entry into force, and currently a revision is under way involving the Commission and a working party of government experts. The Commission will withdraw from the very active position it previously took, while still retaining the final say over criteria. The proposal COM(96)0603 shows very clearly that the Commission's ambition is to facilitate the establishment of a private organisation, the European Ecolabel Organisation (EEO). This organisation will mainly act as the coordinating body between the national competent bodies which will actually run the EU scheme in the future (European Commission 1996a, 1996b, 1996d).

The original idea behind the regulation was that a harmonised market needs to have harmonised market instruments in order to avert trade distortion. As with most official national schemes, however, the EU Flower was not originally designed with sufficient understanding of how to harness market forces successfully. The scheme was voluntary and its objective was only to provide consumers with information, a rather weak ambition from either a market or an environmental-protection point of view.

Besides procedural problems and intra-Commission conflicts over details in criteria approach, the EU scheme has never had any supporters in the market. Efforts to define criteria for textiles, for example, lead to difficulties between Directorate-General (DG) VI and DG XI concerning how the use of pesticides and fungicides should be dealt with (it was pointed out that the draft textile criterion would have excluded from ecolabelled products certain pesticides allowed in agricultural production). Another example concerns that of paper, where the whole scheme itself was threatened when foreign producers pushing third-world governments in front of them claimed that the scheme conflicted with rules on international trade (Vogel 1998).[17]

The EU scheme also has difficulties in attracting supporters in the European market. Producers are not in favour of a label which is not visible and which is largely unknown. Producers are generally very reluctant to succumb to environmental demands on their products and production via market mechanisms until they are forced to by overwhelming demand.[18] The existence of national labels, whether successful or not, reflects the existence of heterogeneous markets. For example, detergents are not identical in northern and southern European markets. Consumers have varying preferences and behaviours, which makes uniform labelling and the use of uniform environmental criteria difficult. It is hard to envisage a future where only very large producers market a few homogeneous products to the entire European market with any success.

For their part, consumers do not know what to demand since there are no ecolabelled products on the shop shelves (and how do you express demand when there is no choice?). In fact, studies of consumer attitudes often find a great willingness to pay a price premium for greener products. However, producers can respond to this preference in several ways. The most common approach is for a producer to attach some green claim to its product, thereby capitalising on consumer preference without actually undertaking environmental changes. Another is to continue to conduct 'business as usual', not reacting at all and leaving consumers without any guidance. A third response is to apply for and adopt an ecolabel.

Demand for ecolabelled products therefore does not become a reality until consumers have proved by their actual purchases that the labelled product is preferred, but demand for a new quality aspect cannot be expressed until the appropriate products are supplied – a real Catch-22 situation. Suppliers have to risk supplying an ecolabelled product, market it so it becomes known to customers, and then hope for a substantial market share. Not until a market share is captured will it be evident to the producers that consumers wanted the product (Plogner 1996).

Moreover, retailers have not been very active on behalf of consumers in terms of promoting ecolabelling by their suppliers. The Swedish example, where retailers take a very active part in promoting the Good Green Buy programme, has not been copied in other European states. There are, however, examples of retailers developing their own generic brands based on environmental-performance criteria, but the criteria are specific to each retail chain and consequently the environmental priorities differ from chain to chain.

Because national competent bodies have invested considerable prestige and financial resources in their own definition of ecolabelling they have a conflict of interest when it comes to promoting an EU label.[19] National schemes are usually funded with the ambition of becoming self-financed via licence fees. Out of these budgets the competent bodies also allocate the resources necessary for work involving the EU Flower. If a national scheme has developed a criterion and attracted licensees to the national label an EU criterion for the same product group might constitute an economic threat. On the other hand, there are financial incentives for member states to propagate the EU Flower: application for the EU label can be made in any country within the Union, but the competent body granting the label receives all the fees based on the applicant's sales throughout the entire European market. There is no mechanism within the EU scheme ensuring that these fees are distributed among the competent bodies.

Environmentalists, consumer representatives and trade unions have not been allowed to influence ecolabel criteria, and have not been given enough resources to examine their quality or assess the idea of promoting ecolabelled products.[20] Therefore these groups are very sceptical of ecolabelling in general and the EU label in particular since this tool appears to provide a new green legitimacy for consumption, totally ignoring the responsibility stated in Agenda 21 (agreed at the Rio Summit) to reduce consumption levels, particularly in OECD countries. Certainly, 'greening' consumption is a necessary element in a more sustainable consumption and production pattern. But if commercially unbiased interests are too weak in the process of defining the goals, 'greening' becomes 'greenwashing', giving a perceived green tone to products or production which in reality undermine a sustainable future. Reluctance to support ecolabelling should thus be seen as a mirror of how unevenly power over the decision-making process is distributed among the stakeholders.

Finally, the EU Flower has had difficulty because the Commission has been heavily lobbied by producer interests with the aim of watering down criteria, obstructing the process and playing directorates off against each other (knowing that in intra-Commission politics the Environment Directorate, DG XI, does not hold the strongest position). In the early months of 1996, for example, American paper producers, in conjunction with third-world representatives, criticised the label as a whole and also attacked it from a World Trade Organisation (WTO) angle by claiming that it was in conflict with the rules of the General Agreement on Tariffs and Trade (GATT) (Vogel 1998).[21] Of course, lobbying is not a new phenomenon, but when incorporated as a normal part of events it is a threat to democracy. The closed, secretive discussions taking place between lobbyists and decision-makers without other interests being able to express their views can never reflect what happens in a democratic and open process. Influence defined solely by financial strength is a recipe for societal disaster.[22]

Improving the design of ecolabels[23]

A market-based instrument like ecolabelling builds on a number of basic assumptions: first, that consumers have a choice in the market; second, that there are consumers with strong preferences for high environmental quality; third, that there are producers willing to supply a higher environmental quality given this consumer demand.

In theory, the market is where supply and demand meet at a price and a quality which buyer and seller are content with. But this theory builds on yet another important assumption, that both sides are equally strong and that knowledge of all existing alternatives is supplied to the market – in short, that the market is not distorted in any sense. In reality, all markets are distorted, competition is never perfect among suppliers or buyers, and consumers are very rarely in a strong position.

Most official schemes have never gone beyond declaring that they serve as a tool for providing information to consumers and have never set environmental targets which market forces should help to accomplish (the need to view market forces merely as instrumental is discussed in Chapter 1). All official schemes are voluntary, meaning that they can never force producers to apply them or consumers to utilise them. In light of this vagueness, perhaps one should not be surprised that most schemes have resulted in labelled products which never really alter existing market balances. Ecolabelled products have captured substantial market shares only in the Swedish market, where, for instance, ecolabelled detergents now constitute 90 per cent of the supplied products. Other claims, whether environmental or health-oriented, have alerted consumers everywhere to act on single issues like dolphin-safe tuna, paper with a high content of recycled fibre, or energy-efficient light bulbs. It is important to remember, however, that these examples are not the product of official ecolabelling schemes. Rather, they prove the failure of most ecolabelling schemes to achieve the same impact as organisation-driven, consumer-oriented consumption campaigns.

Ecolabelling can become an extremely effective tool for changing a market in favour of new production techniques, new product formulations and improved functions. However, in order to achieve this a good portion of market 'muscle' has to be developed by empowering consumers with information on why and how ecolabelling works. Empowerment can be achieved by environmental non-governmental organisations (NGOs) in conjunction with untraditional partners who have substantial market influence, such as retailers, or by boycotting individual producers.

The following three cases taken from the work of the Swedish Society for Nature Conservation (SSNC) prior to the establishment of the Good Green Buy ecolabel scheme,[24] and from subsequent experience with the scheme, highlight these issues of proper and improper instrument design, revealing in more detail how ecolabels can deliver substantial environmental improvement, often alongside economic benefits for the firms involved.

Case study 1: paper

For the last century the paper industry has constituted one of the backbones of the Swedish economy. A major part of the production, around 60 per cent, is exported to European and overseas markets.

SSNC has tried to influence forestry, pulp and paper production in order to save endangered species from extinction. One step in the production process has been especially detrimental to the environment, namely to coastal waters and marine life. The traditional method of bleaching pulp with chlorine has long been criticised by environmentalists. Environmental pressure through public opinion, distribution of information and debates in the media achieved little success. Demands to change production processes were always countered by the industry with arguments like: 'We only produce what the consumers want' (see Plogner 1996). This is of course true, but assumes that the market at the end of the 1970s truly reflected what consumers wanted. In fact, the market was homogeneous – no 'alternative' products were available to consumers and consumers were not aware of the effects of chlorine bleaching on marine life.

From the late 1970s onwards knowledge of the negative environmental impact of the traditional bleaching method slowly spread to larger groups in society, the breakthrough coming in 1986–8, when seals on Sweden's west coast died on a massive scale. The waters were poisoned by blooming algae and life in the sea was threatened. This led to a massive debate on how Sweden took care of the environment and also became the largest political issue of that year's election. A general feeling spread that immediate action must be taken, disrupting the traditional Swedish attitude that authority was doing what was needed.

In 1988 SSNC together with Friends of the Earth Sweden published the second edition of a tiny booklet entitled *Paper and the Environment*, in which environmental criteria for paper were established (they centred on the discharges of adsorbable organic chlorine (AOX) per ton of pulp produced; any method of bleaching was acceptable as long as the discharge of AOX complied with the limit values of the criteria).[25] Also included were lists of products which adhered to the criteria.[26] This action led the organisation of Swedish municipalities to recommend that their members buy their office paper from one particular paper mill in Sweden, whose product adhered to this criterion.[27] The paper mill is small and did not at that time have a market share of any significance. The recommendation clearly threatened the current balance of power in the market.

The combination of straightforward product recommendations to individual consumers and information to large consumers reconfigured the

Swedish paper market in just a year. When 'true' demand was revealed, paper mills were forced to change production processes. On the consumer side, the demand for unbleached paper was manifested most clearly in the market for disposable nappies. The market war became fierce and violent, and 'forced' one of the Swedish pulp mills to become the best in the world when it comes to low discharges of chlorine compounds. A side effect of this was that in a few years this particular mill had more orders than it could meet from the rest of the world, which of course led to a unique position in terms of pricing its product – not a bad side effect of being 'forced' to become market-oriented.[28]

In just two years anything but unbleached or environmentally friendly bleached fibre became impossible to sell on the Swedish market. The result for the environment is that discharges of chlorinated organic compounds from pulp and paper mills have been reduced from 175,000 to less than 1,500 metric tonnes per year.

When the Good Green Buy scheme was established late 1989 the first generation of criteria for paper and pulp were identical to the demands published in the booklet mentioned above. Establishing criteria like the AOX limit was very successful and dealt with what SSNC felt at that time was the top-priority problem when it came to the paper and pulp industry. In order to get the process started SSNC decided to leave other aspects of the production process for the future. Once the problem of chlorine compounds was reduced, other problems arose as important. SSNC's current criteria (generation 4) focus on issues like sustainable forestry management and energy consumption in production, as well as chemical issues.

Case study 2: batteries

In 1989 SSNC produced the first Good Green Buy criteria for AA batteries. The criteria stated that in order to receive the label a battery could have a maximum content of 25 ppm (parts per million) of heavy metal (cadmium (Cd) and mercury (Hg)). It was already known that heavy-metal-free rechargeable batteries existed and one might have suspected that new technology dramatically reducing heavy-metal content also existed for disposable batteries.[29] Sweden had also adopted legislation in 1989 prohibiting the sale of batteries with a heavy-metal content of more than 250 ppm (ten times the hurdle in the SSNC criteria).[30]

When launching the criteria, SSNC bought a range of batteries and analysed their heavy-metal content. By chance, an illegal batch of batteries containing 4,000 ppm was found. The producer happened to be the market leader in Sweden at the time. As a consequence of the publication of the test results in the SSNC magazine, the producer lost a contract with one of the retailing chains, as well as enormous amounts of goodwill (SSNC 1990), but it was also given an enormous incentive to gain back some of the lost goodwill. Within six months it introduced the first heavy-metal-free single-use (not rechargeable) AA battery on the Swedish market. The rest of the producers followed suit within the subsequent six months. New heavy-metal-free batteries were introduced and the old ones were withdrawn from the Swedish market. Since the Swedish market for batteries could probably best be described as an oligopoly, it was not really disturbed. After the initial turbulence the same companies dominated supply, but the product was completely new.

This change would have occurred eventually but it would have taken a much longer time without SSNC action. For example, in 1992, two years later, heavy-metal-free batteries were still nowhere to be seen in Finland. The Finnish competent body, giving its views on proposed criteria for batteries within the Nordic labelling scheme, commented that it had heard of heavy-metal-free batteries so perhaps the criterion could go as far as demanding zero heavy-metal content.

Case study 3: detergents

In 1990, when the SSNC established a collaboration with Swedish retailers, one of the first criteria developed was for detergents. Detergents constitute a major chemical product among those consumed by households. In Sweden, annual consumption is in the magnitude of 50,000 metric tonnes. In order to reduce the environmental burden from detergents, criteria were defined which allowed 1–2 per cent of the products on the market to be labelled.[31]

Needless to say, nothing much really happened. The retailing chains did not have any products to label, the consumers could not find any labelled products in their ordinary shops and the retailers claimed that demand was non-existent. One could see that there was a need to convince one of the major detergent brands on the market to introduce an ecolabelled product. Since retailing is highly centralised in Sweden the selection of products is largely the same across the country. Any producer selling to the major retailers has to be able to supply products, sales support and marketing on a national scale. It was decided that it was necessary to target the largest producer, as small producers, no matter how environmentally friendly their products, face these almost insurmountable barriers to market access. SSNC therefore sent Lever a letter asking it to fulfil its responsibility as the single largest polluter in the detergent market, and pointed out to Lever that by introducing an ecolabelled product it could reduce the environmental impact of certain substances by up to 30 per cent. Lever replied that there was no customer demand for other types of detergents than the ones it already supplied and that they were extremely hard to manufacture without sacrificing product performance.

One month later SSNC asked Swedish consumers to boycott the number-one brand, VIA, manufactured and owned by Lever.[32] Although SSNC normally keeps a rather low profile, the boycott was highly publicised. Furthermore, it alerted consumers to the fact that washing clothes has an effect on the environment. It also gave Lever something to think about since it started losing sales. It took Lever about six months to react to the boycott, at which point it introduced the first multinational ecolabelled detergent on the Swedish market. The previously labelled products were both domestically and foreign produced but none came from a large company. The multinational product quickly became a number-one seller, as SSNC had predicted it would. When our criteria were first published in 1990 the market share for ecolabelled products was hardly measurable. Three years later it was approaching 50 per cent. Currently, the market share of ecolabelled detergents is around 90 per cent.[33]

These examples point out the necessity of combining potential consumer strength with ambitious but feasible criteria. The other very important issue is to motivate public opinion to purchase ecolabelled products once they are on the shelves. Unless producers feel that the label adds to their market edge they will not continue to support it, market it and supply it. Eventually the labelling becomes self-enhancing in a sense. Once it is established, producers strenuously support this quality, and consumers demand it to a higher extent and in a broadening range of products.

The existence of a private Swedish scheme like that of SSNC helped promote the establishment of the official Nordic White Swan programme in Sweden. Also, since the Good Green Buy scheme has been operated in parallel with an ongoing campaign called Shop and Act Green, SSNC has the ability to mobilise active members in consumer campaigns and in campaigns directed at local retailers. The detergent boycott forced the producer to introduce an ecolabelled product (and, incidentally, Lever chose to label its product with the Nordic Swan and not the SSNC label). Once the ice was broken, the other multinationals, Colgate-Palmolive and Proctor & Gamble, soon followed suit. The Nordic Swan has had breakthroughs in those sectors where parallel criteria within the Good Green Buy scheme existed first.

Retailers play an important part in promoting labelled products through their suppliers. Retailers are the real consumers in the market: what they choose is what their customers can choose from. At the same time, retailers have the best contacts with consumers and know first-hand what their preferences are. When they chose to feed these environmental demands back to their suppliers this resulted in labelled products on the shop shelves.

In the other Nordic countries the White Swan is not as common as in Sweden, and no parallel private ecolabelling schemes exist. It is naturally hard to speculate about what the situation in Sweden would have been without the Good Green Buy scheme, but it is likely that Sweden would not have seen market shares of 90 per cent for ecolabelled detergents.

Essential elements in any successful ecolabelling scheme

Based on the evidence of the EU Flower and the three Swedish cases, the final part of this chapter identifies the essential elements in any successful ecolabelling scheme. These elements are grouped under eight headings.

1 Clear ambitions

In order to attract public participation, any scheme should be based on a policy clearly defining its scope, objective and strategy. Ecolabelling must be considered one of the tools for developing a sustainable society, not a universally applicable measure which diminishes the importance of fiscal or legal instruments. A functioning market with ample competition and high consumer participation provides a potential for rapid transition. Markets characterised by monopoly, or oligopoly, demand different strategies, where market pressure has to be utilised in other forms. The strategy of the scheme should also include long-term objectives in order to avoid short-term gains which prove to be long-term mistakes.

In order to fulfil the long-term objective of achieving a sustainable society, any ecolabelling scheme must be able to advocate changes in behaviour as well as changes in products. Buying ecolabelled products is only half the solution, the other half being proper use and reductions in overall consumption levels. Since a tiny label cannot communicate this, the scheme must be supported by consumer or environmental organisations with the objective of educating consumers.

Clear ambitions also apply to the environmental priorities and working principles. In Sweden, legislation concerning chemical substances contains what is called 'the principle of substitution'. In short, the spirit of this principle is that substitution of a substance should occur, regardless of concentration, if alternatives with reduced environmental and health effects are available. This principle should be incorporated into the criteria of all ecolabelling schemes, so that no harmful substances can be accepted, even in very low concentrations, if better substances are available.

2 Independence

In order to gain credibility, which is a prerequisite of consumer acceptance, ecolabelling schemes must be independent with respect to the source of finance and the input of knowledge or information. Any system relying on the financial cooperation of the producers will find it difficult to balance its own aims and the aims of the producers. Most producers oppose any system which disqualifies a majority of the existing products on the market.

Unless the establishment of criteria can be based on reliable, accurate and up-to-date information, the process may become strenuous and erratic. A lot of competence is naturally found among producers, and their participation during the drafting stages prevents obvious mistakes. The best method is to find a scheme which affects the producer more seriously if the criteria are ill formulated and erratic than if they are based on accurate information.

Independence does not mean that criteria are neutral. Criteria are formulated to promote excellence and are based on the assumption that visibility is a necessary prerequisite of success. Naturally, this favours currently marketed products which already deliver excellence. Financial independence means that the establishment of criteria should not be held hostage by those whose commercial interests depend on the outcome. In the case of SSNC, financial support from retailers is structured so that resources come out of overall sales profits, not only from sales of labelled products. In order to guarantee that retailers cannot exercise undue influence over criteria formulation, SSNC has the sole right to make the final decision on all criteria. Since it is not in SSNC's interest to perform criteria formulation based on erratic or biased information, the process leading up to the selection of a criterion involves several hearing stages in which as many stakeholders as possible are invited and contacted. SSNC risks its entire credibility every time a criterion is published. Credibility is what draws in society members, who in turn provide the mandate to act for the environment, not only in ecolabelling issues but also in all other aspects of SSNC's work.

3 Non-discrimination

The system must be accessible to all producers, regardless of their size. Most small producers are more motivated to adhere to new criteria than large ones. This can be attributed partly to the difficulty of gaining access to the market unless the product is helped by a unique quality. The ideal scheme would be designed in such a way that it does not discriminate amongst producers on the basis of size or financial strength. Most official schemes are financed by taxing participating products through licensing fees or turnover-related fees. This constitutes a cost burden on desirable products. Instead, products not complying with environmental criteria should face an environmental tax in order to offset their negative environmental effects.

4 Maximum market impact

In achieving ambitious environmental objectives, ecolabelling schemes must formulate criteria which take into account the actual market situation. Maximum impact is a consequence of the speed at which new improved products can be introduced, as well as of the ease with which it is possible to reformulate existing products according to the demands put forward in the criteria. The same effect can be achieved by placing extremely strict demands on a few items or from small reductions on large numbers of items. But effective schemes never attempt to place labels on all products in the market at the same time. Equally important, there is little scope for improving an environmental situation if the criteria established depict some 'best' product still on a drawing board.

5 Consumer participation

Especially in the initial stages, consumer demand has to be organised and catalysed. Participation of consumers in this context involves any party acting as a consumer, the larger the better. One such example is government- or state-owned enterprises, which can specify the products desired through large-scale procurement policies. Producers then have to choose between being able to supply, and therefore having to meet the demands, and forfeiting this large market segment.

Individual consumers are traditionally very weak, especially if the market consists of monopolies or oligopolies. This was certainly the situation in Sweden when SSNC started its own scheme. Consumers were environmentally concerned, but the highly standardised assortment of available products offered little scope for manifesting this demand. If consumer voices had not been organised and heard very little would have happened.

It now appears that the strategy among the large producers has changed. Instead of neglecting these demands, they now all strive to adhere to the criteria in order to avoid a competitive disadvantage. If this situation is prolonged, the products are environmentally improved but the development process is halted. To move beyond this plateau, consumers have to be informed and remobilised. To do this, both environmental and consumer organisations with high credibility and good channels to consumers are needed.

This also implies that, as a prerequisite of their success, consumer and environmental organisations should be well represented and have strong influence on the boards or steering committees of ecolabelling schemes. Unfortunately, in the official systems currently operating the opposite is often the case.[34] This is a consequence of many factors; in particular, their lack of funds undermines the ability of these organisations to devote personnel to time-consuming work in subcommittees and expert working groups.

Judging from experience, any system influenced by producer interests has a hard time installing schemes or criteria that will actually have an effect on the market. It is impossible for producer organisations to combine protecting the financial interests of their members with participation in a scheme where 80 or 90 per cent of their members may be disqualified. Thus the high aspirations of consensus models are in vain.

6 *Transparency*

In order to avoid criticism, ecolabelling schemes must be transparent. The reports or investigations constituting the basis of the criteria document should be public. Openness is a prerequisite of long-term credibility because it makes external reviews of the work possible. Another important aspect of this approach is that unpublished data from producers cannot be accepted as a basis for criteria. Only by publishing their findings and expanding the available knowledge can producers legitimately influence the process. The means by which to construct transparency can vary. In the system designed by SSNC, a transparent and strict method has been applied. Actual criteria formulation consists of a number of 'hurdles' which must be jumped by the producer interested in attaching the ecolabel to its product. SSNC has also been working with two-level criteria which give advance notice of what the next step will be in the revision of the criteria. This gives producers a goal when considering reformulating or redesigning their products.

The other type of criteria commonly used, discussed previously as a matrix, is preferred by producers because it introduces greater flexibility by allowing trade-offs between parameters. However, this system sacrifices transparency, as it is almost impossible to gain access to the specific environmental merits of a certain labelled product unless one is actually on the evaluation board.

7 *Cost efficiency*

If a scheme has high costs and low output, the dynamics of the system are lost to the detriment of the environment. This disqualifies strict life-cycle assessment methodology as a basis for criteria formulation. As discussed earlier, designing and evaluating LCA is a process beset by disagreements. Waiting for this process to reach a consensus might involve lengthy delays or create permanent policy-making paralysis, leaving consumers with no guidance whatsoever.[35]

Cost-efficiency, needless to say, implies that a large bureaucracy should be avoided. The SSNC system has been able to do this, producing criteria at approximately 25 per cent of the cost incurred by the Nordic scheme for corresponding criteria.

8 *National rather than international action*

Developing an optimal national system which aims to diminish the environmental impact of consumer products is rather hard to combine with the ever-increasing internationalisation of trade. In the long run it is reasonable to assume that the technical structure will converge. In the meantime it is better to regard ecolabelling as a rather local activity, and at best strive for some international consensus concerning the basic goals. One way could be to incorporate international agreements, for example those on the reduction of emissions of carbon dioxide (CO_2) and chlorofluorocarbons (CFCs).

Any market-based instrument needs to be developed with a thorough understanding of how the market in which it exists functions. If, for example, all EU consumers have the same preferences concerning product performance and how products are used, uniform ecolabelling criteria are logical. If, on the other hand, consumers have differing preferences and use products differently, ecolabelling must adapt to its target consumer group, be it local, national or international. Since ecolabelling is voluntary from both a consumption and a production point of view, it needs to attract consumer acceptance by being perceived as logical and to the point.[36] Thus a basic set of common EU parameters can be the basis for national labels where additional national criteria are accepted. In the long run this will permit a convergence of criteria when preferences and environmental priorities converge, but in the short run it is 'think global, act local' which should guide the design of ecolabel programmes.

Conclusion

Because it is a market-steering mechanism, in order to be successful an ecolabel has to be situated in a functioning market. New alternative products must be able to gain access to the market, and all agents in the market – producers as well as consumers – must be able to get information.

Good ecolabelling concentrates on factors which are logical to the consumer. Any consumer would like to understand why his or her purchase improves the situation and why the previous choice was detrimental. The long-term objective of ecolabelling is to educate consumers in order to give them knowledge of the environmental effects of consumption and to give them sufficient strength to be able to make informed purchase decisions, thereby maximising market efficiency. If criteria are set at a level which leaves a sufficient percentage of the market above the limit, then demand can shift consumption patterns as traditional products lose market share.

Ecolabelling and market tools in general depend upon the willingness and ability of consumers to accept the role of agents partly responsible for countering environmental impacts. Still, consumers need to be empowered if they are to be able to do this at the same time as governments are abandoning traditional protective legislative or fiscal measures.

In the case of the EU these issues are readily apparent. Completing the single market and expanding trade are the objectives, and this leads to increased environmental degradation. The current withdrawal from corrective measures like legislation or fiscal instruments (for example the CO_2 tax), and the increased reliance on market instruments and voluntary agreements is a dubious means of achieving the environmental goals previously agreed upon. But whether ecolabelling represents a panacea or a Pandora's box cannot really be answered yet, as it all depends how the Commission drives the revision of the regulation.

For instance, the Commission could utilise its own buying power and buy only products that adhere to the criteria developed within the EU Flower scheme and other ecolabelling schemes. Current producer resistance to apply for the label would only result in the loss of a huge customer. Unless the Commission tries to adopt a much more dynamic mode of operation the EU scheme will never become more then a tiny niche; national official schemes will struggle on at a slightly higher impact level and private schemes like the one SSNC operates will be the exception to the rule – that ecolabels are not tools for change.

Notes

1 The Club de Bruxelles provides a good account of the growing interest in measures aimed at encouraging rather than enforcing good environmental behaviour (Club de Bruxelles 1995).

2 Private ecolabelling schemes are not operated in conjunction with a national government but by private organisations such as the Swedish Society for Nature Conservation. In contrast, 'official schemes' have a relationship with government, either financially, by being operated jointly or by being installed through legislation.

3 Regulation 880/92 states that the Commission must consult interest groups on the proposals for criteria developed within the EU scheme. A forum has therefore been assembled consisting of three representatives each of industry, consumers, environment and trade. Trade unions were inexplicably excluded and participate on an unclear mandate. The forum has no real power but is allowed to register its opinion once for each criteria document.

4 Extended producer responsibility is inspired by the German *Verpackungsverordnung*. The Swedish model builds on discussions, sector by sector, between government and industry, in which targets and dates are established for waste reduction/recycling. The sector in question then accepts the responsibility of achieving the targets and developing the means to do so without strict prescriptive legislation stating how it shall implement it (for discussion of negotiated agreements, see Chapters 4 and 8).

5 On the basis of the 'new approach' established by Directive (83/189/EEC), the legislative authorities (the European Parliament and the Council of Ministers) have recourse to private standardisation through the European Committee for Standardisation (CEN).

6 Standardisation has become an intrinsic part of modern society. As one study notes:

> Almost all industrially manufactured things around us have been standardised in some way or another. In day-to-day practice, it is not so much the stipulations of law rather than these very technical standards that determine how a product is made and in what manner polluting facilities are operated. . . . But the process of standardisation has until now maintained a very low profile, not to say a secretiveness – it has at

> all events eluded any broad public participation. The actors in this process keep to themselves. Only the lobbyists with a vested interest in a specific project have a say.
>
> (Führ *et al.* 1995: 3)

7 Within the ISO 14000 series a number of standards are currently being developed which deal with ecolabelling and life-cycle assessment (LCA).

8 One example is detergents, where surfactants constitute one of the basic concerns. The Good Green Buy scheme has defined demands on degradeability and toxicity for surfactants based on numerical hurdle values. The criteria include a list of the surfactants which, according to publicised data, fulfil these values. The list contains six groups of surfactants, where group 1 contains those with the best overall environmental performance. The criteria go on to define groups 1–3 as acceptable in an ecolabelled product. Any producer can find out how its product performs in this respect. This principle is then carried out for other functional groups or substances in the product. For a complete description of how the Good Green Buy scheme operates, see Eiderström 1997.

9 The formulation in Council Regulation 880/92 (OJL99, 11 April 1992) on a Community ecolabel award scheme is: 'The specific ecological criteria for each product group shall be established using a "cradle to grave" approach' (Article 5: 4).

10 For a summary of discussions held at the WTO Committee on Trade and Environment meeting of 26–27 October 1995 on the 'widely different LCA methodologies' forming the basis for ecolabelling criteria, see World Trade Organisation 1995.

11 Discussion has just begun on how to incorporate accurate assumptions into the models, for example in the shape of life-cycle stressor-effects assessment (LCSEA). As an example of the difficulties with, for example, electricity, see Groupe des Sages 1995.

12 One study describes the situation, as follows:

> Experience has shown that particularly at the European level lobbyists use the standards process to exert political influence upon EC legislation. . . . That this process permits lobbying at all is partly because standardisation has long been defined and accepted as being part of self-regulation of industry.
>
> (Führ *et al.* 1995: 3)

Council Regulation 880/92 (OJL99, 11 April 1992) stipulates that the Commission shall 'consult the principal interest groups who shall meet for this purpose within a consultation forum' (article 6: 1). It does not define what status is accorded to the outcome of the consultations.

13 Within ISO work is under way on a compliance-assessment procedure which guarantees that ecolabelling programmes adhering to the forthcoming ISO standard utilise reproducible methods to set their criteria. It is not known to the author whether any of the current ecolabelling schemes contain a formal appeals process. Currently, there is a conflict between the Swedish board for the Nordic White Swan and one of the detergent manufacturers concerning the decision to deny a product the Nordic label following a change in the criterion for detergents. The granting of labels is not a transparent process in any scheme. All data concerning an applicant's product are kept secret for commercial competition

reasons and the only party able to dispute the decision on a specific product is the applicant.

14 Initially, ecolabelled products are priced higher than conventional products. In the long run, as has been seen, for example, in Sweden with detergents, the price of the ecolabelled product does not vary from that of its conventional competitors. The reason for this is probably that during the first stages of labelling the customers are few and place a high priority on choosing the environmental quality at any cost. As the market share grows, the targeted customer group still views environmental quality as important but 'incorporates' it into an overall expectation about the supplied product's quality, so that price once again becomes the basis of competition.

15 Regulation 880/92 (OJL99, 11 April 1992) states that the principal interest groups (represented at community level) shall be consulted and allowed to deliver their opinion on criteria proposals prior to the Commission presenting the final proposal to the Regulatory Committee.

16 There is no formal explanation of why this change in approach was chosen by the Commission. A note dated 13 December 1995 states that the chosen approach is well in accordance with the existing procedural guidelines, which the Commission describes as 'informal'. But the note makes clear that the Commission 'wants to exert a closer control over compliance with methodological requirements, completeness, transparency and neutrality of the study' (European Commission 1995).

17 In mid-1996 there were lengthy debates on the criterion for fine paper where, representatives of the American Forestry and Paper Association and representatives from Brazilian paper producers expressed their opinions on the proposed criteria.

18 This statement is supported by the findings of Ann-Charlotte Plogner, who has studied what happened in Sweden when ecolabelling was introduced for detergents. She concludes that the existence of what she calls 'industry logic' prevents industry from recognising the development of new markets and delays their adaptation to, for example, ecolabelling (Plogner 1996).

19 As one proposal argues:

> There is no reason for Sweden and the Nordic countries to quietly await the common co-ordinating work in Europe aimed at developing a functional ecolabelling. Rather, for environmental reasons, the work and experiences accrued from working with the Nordic White Swan should be aggressively exported to other countries. . . . It could well be that the Nordic White Swan develops into a 'premium brand', the golden logo, the best in show-logo, with the highest environmental demands, accompanied by, for example, the EU Flower or another national logo representing 'acceptable according to least common environmental denominator (least environmental demand in order to be accepted on shop shelves)'. The White Swan has all the prerequisites to develop into the party carrying the yellow leader shirt.
>
> (Lighthouse 1995: 22)

A more sanguine view of the EU ecolabel was expressed by the working group representing the French Ecolabelling Board in the context of proposals for a revision of Council Regulation 880/92:

> Concerning relationships with national ecolabels, proposed lines are welcomed by the group because:
>
> - they show that the Commission recognises existence of national ecolabels, and the fact that national and European ecolabels may introduce a synergy on the market, the one pushing the other.
> - they may be a solution to the need for flexibility expressed through the 'graduation' proposal.
> - lastly, they [reflect] the fact that European criteria express a compromise between 15 countries. In complement, national ecolabels may address national criteria.
>
> (French Ecolabelling Board 1996: point 9)

20 Article 6 of Regulation 880/92 (OJL99, 11 April 1992) describes how the consultation of interest groups should proceed:

> The Forum should involve at least the Community-level representatives of the following interest groups:
>
> - industry (including trade unions as appropriate)
> - commerce (including trade unions as appropriate)
> - consumer organisations
> - environmental organisations.

Each of them may be represented by having a maximum of three seats. The participating groups should ensure appropriate representation according to the product groups concerned and having regard to the need to ensure continuity in the work of the consultation forum.

The Regulation provides for access but does not state that the forum has any formal mandate to intervene or exert any influence on criteria formulation, nor does the Regulation ensure that financial resources are made available to provide interest groups with equal opportunities.

21 The US generally expresses very strong feelings when it comes to environmental demands across borders. The EU Committee of the American Chamber of Commerce in Belgium has stated that 'The EU Committee opposes governments making judgements on imported products on the basis of the PPM's (production methods) used to manufacture them. . . . Trade measures should not be a tool of first resort to address global and regional problems' (Amcham 1996: 3, 7). The EU Committee also noted that the EU ecolabelling scheme had been placed on the US trade barrier 'watch list', and argued that 'if ecolabels are allowed, their criteria should be harmonised internationally and they should be information-oriented (e.g. nutrition labels) so that any producer could supply them' (*ibid.*: 2).

22 As a Swedish newspaper noted about lobbying within the EU, 'Interest organisations for everything from candy and toys to weapons and cigarettes mingle in the corridors of Brussels' (*Dagens Nyheter*, 21 October 1996). (See also Dawkins 1995.)

23 This section is adapted from Eiderström (1997).

24 The Good Green Buy scheme was developed by the Swedish Society for Nature Conservation (SSNC) in collaboration with three Swedish retailing chains. In many respects it differs in approach from official schemes described in this chapter. One of the main differences is that an environmental organisation has the ultimate say when criteria are defined.

The process leading up to criteria establishment is open to all interested parties and usually involves open hearings in order to minimise the possibility of individual producers' aims exercising undue influence. All information upon which decision-making is based has to be published and subsequently evaluated within the scientific community in order to be considered, a requirement which also helps disseminate available alternative product ingredients and technology. The criteria are usually based on a hurdle system, incorporating the principle of substitution (discussed at the end of this chapter). (See also Eiderström 1997.)

25 When the Swedish environmental movement published the first edition in 1987, with the title *Unbleached for the Sake of the Environment*, it represented a major shift in strategy since all previous action had been directed specifically towards industry and legislators, without much success.

26 It is important to note that this example shows what SSNC did prior to the establishment of the Good Green Buy scheme. It was, however, how SSNC came to understand the power of mobilised consumers.

27 The paper mill is Munkedal, a tiny mill on Sweden's west coast.

28 One indication of the impact of environmental demands put forward by ecolabelling is that environmental investments as a share of total investments during the period 1988–92 rose from 13 per cent to 29 per cent. According to Göran Phorse of Pappersgruppen, ecolabelling was the major reason for this (Kronbladh and Lagerstedt 1995).

29 As early as 1988, when the author was researching the battery industry for new environmentally benign alternatives, material from VARTA described new rechargeable batteries based on NiH-technology. They were not introduced in any market, however, which seemed to be a case of 'cash-cow' mentality – when a product has reached maturity and is paying back maximally to the investor, thereby discouraging the introduction of newer products which generate lower profit per unit.

30 The SSNC criteria were not legally binding since they were defined within a voluntary ecolabelling scheme.

31 The hurdles in the criterion were defined according to environmental properties of substances used in the products, but since this was an initial stage in defining criteria for the product group, knowledge of the actual market composition was rudimentary in terms of what the products supplied were composed of. Aiming at any degree of market share therefore involved a measure of guessing.

32 SSNC and the Shop and Act Green Campaign campaign every year on green consumption issues during an Environment Friendly Week. Two weeks before this campaign in 1991 a brochure was distributed to the active members of SSNC, giving them the rationale behind the boycott and also instructing them on what to argue when talking to the public. SSNC local societies then incorporated the boycott in their Environment Friendly Week campaign strategy, which gave it wide notoriety.

33 Since 1992, when multinational resistance to ecolabelling was broken, the composition and strength of the producers on the market has changed, not so much as a consequence of ecolabelling as such, but rather in conjunction with the trend of generic brands capturing larger segments of the market since the retailers promote these in favour of brands from outside suppliers. As a consequence, the market share of multinational products has probably diminished compared to that of the generic products (which are all based on environmental strategies).

34 The drafts presented at the second meeting of the Working Group of Government Experts on the Review and Revision of Regulation 880/92 show

that even the EU scheme will completely marginalise interest groups by eliminating the Consultation Forum when the EEO is established (European Commission 1996a, confirmed by European Commission 1996d). When formal consultation is not stipulated, informal consultation and lobbying take place and, as was discussed earlier, are dominated by commercial strength.

35 In December 1993 the European Commission set up the Groupe des Sages (GdS) to advise on the role of LCA in the EU ecolabelling programme. The GdS met four times and produced a first report in September 1994. The GdS concluded that LCA can make a significant contribution in providing a scientific, unifying and transparent basis for the EU ecolabelling programme. At the same time it was concluded that LCA is still a developing methodology, requiring additional research and systematic data collection to improve its application. In the long run the results of this research should also be brought in line with ISO standardisation (Groupe des Sages 1995).

36 Within the EU scheme the question of water use has been raised by Spain, where shortage of water is a pressing issue. Consequently, the Spanish would like to promote water reduction in any criterion where this is possible, whereas other member states attribute much less importance to this issue. Spanish consumers might view the omission of water efficiency as particularly illogical, and might prefer products which highlight this concern over EU-labelled ones which do not.

References

Amcham (1996) *EU Committee Position Paper on International Trade and the Environment* (Brussels: American Chamber of Commerce).

Club de Bruxelles (1995) *New Policies and the Greening of Industry* (Brussels: Club de Bruxelles).

Dawkins (1995) *Ecolabelling: Consumers Right to Know or Restrictive Business Practice?* (Minneapolis: Institute for Agriculture and Trade Policy).

European Commission (1995) Document XI.E.4/BD/ty D(95).

—— (1996a) *Consultation Document on the Revision of Council Regulation No. 880/92 of 23 March 1992 on a Community Ecolabel Award Scheme*, Commission of the European Communities, 25 June.

—— (1996b) *EEO, European Ecolabel Organisation, the Statutes of EEO – Draft Proposal, Option 1, CEN Model*, Commission of the European Communities, 28 May.

—— (1996c) *European Commission Communication to the Council and to the Parliament on Trade and Environment*, COM(96)54, 28 February.

—— (1996d) *European Draft Proposal for a Council Regulation Establishing a Revised Community Ecolabel Award Scheme,* COM(96)0603.

Eiderström, E. (1997) 'Swedish shoppers seek "green" label', *FORUM for Applied Research and Public Policy* 12(1): 141–4.

French Ecolabelling Board (1996) 'Position paper', 12 June, on file with author.

Führ, M., Brendle, U., Gebers, B. and Roller, G. (1995) 'Reform of European standardisation procedures, requirements of constitutional and European law upon standardisation through private bodies', Department of Social and Cultural Sciences, University of Darmstadt, Darmstadt, on file with author.

Golub, J. (ed.) (1998) *Global Competition and EU Environmental Policy* (London: Routledge).

Groupe des Sages (1995) 'Research needs in life-cycle assessment for the EU ecolabelling programme', Leiden, July, on file with author.

Kronbladh, E. and Lagerstedt, A. (1995) 'Ecolabelling and its impact on the fine paper industry', Gothenburg School of Economics, mimeo.

Lighthouse, A. B. (1995) *Miljömärkning i Sverige: ett Strategiförslag* [Ecolabelling in Sweden: a strategy proposal] (Göran Gennvi).

OECD (1995) *Joint Session of Trade and Environment Experts, Summary Record of the NGO Workshop*, 18 December (Paris: OECD).

Plogner, A. (1996) *Miljöanpassning och Strategisk Förändring: En Explorativ Fallstudie om Branschlogik, Tvättmedel och Miljö*, Swedish Environmental Protection Agency, AFR-report No. 132 (contains summary in English).

SSNC (1990) *Sveriges Natur* 1.

UNCTAD (1995) *Final Report of the Ad Hoc Working Group on Trade, Environment and Development*, 10 November.

Vogel, D. (1998) 'EU environmental policy and the GATT/WTO', in J. Golub (ed.) *Global Competition and EU Environmental Policy* (London: Routledge).

World Trade Organisation (1995) 'Trade and Environment', 8 December.

10

ENVIRONMENTAL MANAGEMENT SYSTEMS: THE EUROPEAN REGULATION

Karola Taschner

Introduction

The Fifth Environmental Action Programme (EAP) aimed at broadening the range of EU policy instruments, because the implementation and enforcement of Community laws had fallen short in the member states. The EAP mentions environmental and audit systems among those instruments. In 1993 the Council of Ministers adopted a Regulation allowing voluntary participation by companies in the industrial sector in a Community eco-management and audit scheme, called EMAS, a programme which has created a stir ever since its creation (EC 1993a). This chapter examines the development of EMAS, compares it to competing international standards and offers an assessment of its merits as a new tool for EU environmental protection.

EMAS is a promising instrument which suffers and benefits from the legacy of the past. It is ultimately a response to the worldwide legislation on producers' liability for damages caused by their products. Such legislation is in place in twenty-five countries and covers approximately between 60 and 70 per cent of the world population. This legislation forced enterprises, for their own sake, to introduce quality-management systems to guarantee the safety of their products. The international standard ISO 9000 series lays down requirements for quality management systems for worldwide use.

The other strand of the past of EMAS is US legislation on environmental liability, which pushed manufacturers to see to it that their environmental risks are as small as possible, i.e. they have – again for their own sake – to strive to improve their environmental performance. Environmental risks are very high and insurance companies were not willing to cover them.

The question whether the implementation of the current international standard (ISO 14001) or a similar standard actually reduces the environmental risks to any measurable degree is still under discussion.

It is by pure self-interest that any US producer will try to minimise environmental impacts since the producer must always be aware that litigation could be started against them. This is, however, also the reason why US manufacturers wanted to have as few commitments as possible written into an international standard, because they feared it could be used against them.

The first standards for environmental management systems were set up in the UK as a small annexe to the quality-management system and developed into the BS 7750 standard in 1992 and, a little later, in 1993 – more gloriously – into the EMAS Regulation.

What is EMAS?

EMAS is a voluntary scheme and participants are industrial sites which want to improve and publicise their environmental performance. For this purpose they have to adopt a company environmental policy containing commitments aimed at reasonable continual environmental improvement. The process sets off with an initial environmental review of the environmental effects of the site (in essence a small environmental impact assessment). Then an environmental programme and an environmental management system (EMS) have to be set up in accordance with the results of the initial review. To check the effectiveness of the system it must be audited. Any audit is of use for the site manager in the first place, but the EMAS Regulation goes one step further and asks for a validated public environmental statement, which, when presented to the national competent body, gives the site the right to registration and the right to bear a label. Environmental policy, review, programme, management system and audit have to meet rather comprehensive requirements which are laid down in the annexes and present the core of EMAS. Though EMAS is a voluntary scheme, sites wanting to be registered have to fill in its requirements to the letter.

EMAS is an attempt to move away from the end-of-the-pipe philosophy with respect to pollution reduction and prevention, and it encourages the design of production processes which take account of the environment right from the beginning. Production processes are screened and optimised for the environment. This in itself is a fruitful exercise since it will help to save resources and reduce the cost of waste elimination. Moreover, management systems result in cost savings through better knowledge of material flows. Since most enterprises have a quality-management system in place for product-liability reasons they have only to adapt this for environmental purposes, rather than starting with an entirely new philosophy.

The public statement is most promising since it could make production processes at the sites transparent and would not leave stakeholders alone with glossy brochures containing incomparable and uncertain data. Reporting to the public in the EU is still in an embryonic stage compared to the US, where emissions have to be reported and the public has a 'right to know', and where not only public authorities but also enterprises themselves are obliged to provide data on request.

The most attractive element of EMAS is its dynamics. All things being equal, the results from the application of EMAS will vary considerably amongst sites because of the different levels from which they start and the different pace at which they will proceed for their 'continuous improvement', which they themselves decide. A site of good environmental performance and a comparatively 'dirty' site can both get registered. This is only acceptable because the 'dirty' site will also have committed itself to the improvement of environmental performance and legal compliance. EMAS sites do not guarantee to meet all environmental demands immediately, but they have signed up to do so in the future, though they will determine the date themselves.

Environmental performance will become a matter for the board of directors and should become, in principle, a matter for employees as well, so the process can develop momentum and involve *people*. Since EMAS involves all personnel, is linked to the top management and also involves employees, those on the spot have an opportunity to contribute to the process and invent schemes for less pollution, less waste generation and less resource use. EMAS could help to build up a corporate identity around the environment and could make people proud of their achievements, allowing room for imagination and inventiveness; enterprises should award efforts for improvement by their personnel.

As discussed in more detail below, Article 12 of the Regulation allows the possibility of crediting national and international standards. Sites can be registered under EMAS when they have been audited according to another standard and then filled in ('bridged') the additional EMAS requirements.

The EMAS procedure

The purpose of EMAS is the registration of an industrial site as fulfilling certain environmental requirements. Sites eligible for EMAS registration are those operative in energy generation, waste treatment and the industrial sectors which are mentioned in section C and D of the so-called NACE index, an inventory of economic activities set up by the Community originally for statistical purposes (EC 1990). Registration concerns industrial sites and not the whole company.

As a strictly voluntary scheme, a company willing to have a site registered under EMAS has usually decided to face up to the environmental challenge. So it is only logical that EMAS demands a decision on environmental policy at the highest corporate level. The environmental policy is then made accessible to the employees and the public. The company commits itself to integrate mitigation efforts into its environmental programmes, management systems and audits. Prevention of pollution, saving resources and certain aspects of protection at the workplace are the principal aims. Furthermore, it commits itself to abiding by a number of defined practices of good environmental management. Article 3a of the Regulation stipulates that a company's environmental policy must contain a provision requiring legal compliance by the site and 'commitments aimed at the reasonable continuous improvement of environmental performance, with a view to reducing environmental impacts to levels not exceeding those corresponding to the economically viable application of best available technology' (EC 1993a).

A site has to conduct an initial review to identify its environmental impacts. This review is the basis for the objectives and targets the company sets itself for environmental improvement. The environmental programme has to be developed accordingly – bearing in mind the reduction of environmental pollution and saving of resources, provision of timescales, supply of human and financial resources, measurement plans and equipment, documentation, etc. – in order to establish a functioning environmental management system.

A representative of the highest management level has to be in charge of the environmental management system. Responsibility, authority and interrelations must be clearly attributed. The environmentally relevant activities of the site have to be documented and monitored regularly. This also includes the pursuit of working operations, training, as well as providing information about the environmental performance of the site to both internal and external parties.

The site has continuously to monitor the good functioning of the environmental management system using internal qualified personnel. After three years at the latest a external person or organisation has to verify the site. The accredited, independent verifier assesses the audit of the environmental policy, programme, management and environmental declaration for their compliance with the requirements of the Regulation.

An essential result of the environmental verification is the public statement, which is written in a comprehensible form to provide the public with information on the site's activities, a description of their relevance for the environment, the site's environmental policy, programme and management systems, emission data in aggregated form, waste generation, resource use, energy and water consumption, and other environmentally relevant aspects and factors. The date of the next public statement and the name of the verifier must also be communicated. Each year, between the verifications, the sites have to draw up simplified environmental statements in which they summarise environmental data and indicate possible changes.

When the site has met all these requirements it can be registered by the national competent body, which has to notify the regulatory authority of this event beforehand in order to allow the authority to report back if the site is in breach of permits or of environmental laws. The site can be registered only when the breach is rectified. The Regulation provides the competent body with the power to delete sites from the register if they are no longer in compliance with the requirements of the Regulation. The competent bodies publish the registered sites and transmit this list to the Commission. Member states have to take care that companies and the public are informed of EMAS, and that the two competent bodies – one for the accreditation of verifiers, the other for the registration – are set up, and are independent and neutral.

The Commission is supported by a Committee (Article 19 Committee) which is composed of representatives from member states and chaired by the Commission. The Commission has invited representatives from interested societal groups: industry, trade unions, environmental organisations.

Since late 1997 the Commission has been consulting with member states and interest groups on revision of the Regulation. The revision endeavours to get EMAS into line with ISO 14001. For instance, Annexe 1 of the current Regulation will be replaced to a large extent by parts of ISO 14001. Only a few elements of the EMAS Regulation which are deemed non-equivalent by the member states (in the Article 19 Committee) are taken up separately. The main EMAS elements are to be introduced into the main text. The scope of the Regulation will be opened, accordingly, to that of the ISO standard, to include organisations (not sites), services, products.

Environmentalists had hoped for a more demanding EMAS Regulation. They will be disappointed because the trend is in the opposite direction, i.e. to apply the more general and flexible wording of the international standard.

Development of the EMAS Regulation

The process of preparing the EMAS Regulation was not easy. One has to be aware of this to understand its design. Acceptance by industry was crucial because without the involvement of the single enterprises the special advantages of EMAS cannot be realised. Therefore member states were eager to act in agreement with industry.

At the time, industry's reluctance to have the EU Regulation adopted could only be overcome by making participation in the scheme voluntary instead of mandatory. Member states tended to voice the reservations of industries which did not favour the adoption of EMAS. Industry favoured an international standard – similar to the British standard – which was in the process of being developed by the International Standards Organisation (ISO). The text of the Regulation has been subject to many changes in the Environmental Council.

The status of some of the key elements in Commission and Council texts were changed or introduced during the negotiations (see Table 10.1):

- Legal compliance was not contained in the Commission Proposals but was obviously a demand from some member states (EC 1991, 1993b). Since the wording became less stringent between the Discussion Document of December 1992 and the final Regulation, there must have been considerable opposition inside the Council against demanding compliance right away.
- Economically viable application of best available technology (EVABAT)[1] seems to have been a German demand, since the Document of December 1992 marks a reservation of the German delegation which was eventually lifted when this new element was introduced into the Regulation – reluctantly accepted by the other member states, given the number of qualifiers. In principle, continuous improvement of environmental performance is limited by setting the goal of EVABAT.
- Mandates to the Comité Européen de Normalisation (CEN) and the European Accreditation of Certification (EAC) were envisaged by the Commission and appear as a declaration to the minutes in the Council Document of December 1992 (EC 1992).[2] A declaration about these mandates appears in the minutes from the final adoption. (Declarations to the minutes are not binding; they are, however, a moral obligation for the Commission.) This is the reason why the Commission acted as if the mandates had been fixed in the text.
- Industry's demand to allow other standards to be declared in correspondence with the requirements of the Regulation was not originally intended by the Commission but appears to have been introduced by the member states.

Table 10.1 Development of the EMAS Regulation

Key provisions	*Draft proposals*			
	I (EC 1991)	*II (EC 1992)*	*III (EC 1993b)*	*IV (EC 1993a)*
Legal compliance	-	+	-	'provision for . . . '
EVABAT	-	-	-	'commitment to . . . '
Mandate for CEN	+	footnote, minuted	+	declaration to the minutes
Correspondence	-	+	+	+
Mandate for EAC	+	footnote, minuted	+	declaration to the minutes

Interests of industry, public authorities and non-governmental organisations (NGOs)

Some companies have been setting up environmental management and audit systems for their own sake to make sure that they do not run the risk of causing environmental damage for which they would be held liable. For them EMAS offers an opportunity to make their achievements known to the outside world in order to improve their public image. For enterprises which have faced public criticism over their environmental performance it will be valuable to demonstrate concern for the environment. Companies will also gain acceptance by their customers, local community, pressure groups, shareholders, banks, insurance companies, regulatory authorities and the public at large, which is not an insignificant matter and represents significant monetary value.

The real motives for participation are improved public image and pressure from the customers. Also, internal advantages have proved to be worth the effort of participating. Large companies often require EMAS participation from their small and medium-sized enterprise suppliers (TEKES 1994). German car manufacturers, for example, welcome EMAS participation by their suppliers, although they do not go as far as Volvo, which made it a condition that suppliers be registered with EMAS by mid-1997 (OB 1996a).

Industry's interest in certification/verification would be greater if it offered them the possibility of self-control, less control by their regulatory authority and deregulation in the long term. But this is still an issue for debate. Member states still have to be convinced that independent verifiers can take over surveillance tasks. EMAS could provide companies with flexibility in dealing with their environmental problems, including the ability to integrate environmental management requirements into their investment cycles.

The regulatory authorities have control over undue EMAS registrations because they have to be notified in advance of a registration and can then report to the registration body breaches of environmental laws by the site. But for the time being local authorities are still hesitant to embrace EMAS.

The position of environmental organisations is ambiguous: on the one hand, they welcome any voluntary attempt for more environmental protection made by industry; on the other hand, they see many loopholes and possibilities for fraud, and are afraid that EMAS might develop into an instrument which will be used instead of and not in addition to public-authority control.

EMAS in relation to standards

Article 12 of the Regulation allows sites to be registered which have certified their eco-management and audit systems according to a standard. Before this can happen the diverse array of national standards and procedures, as well as the emerging international standards, have to be recognised by the Commission acting in agreement with the Article 19 Committee. Article 12 was added to the Regulation following demands from industry. While the Regulation was under negotiation in Council and Parliament, industry had even asked that any certificate based on one of the many possible standards be declared compatible with an EMAS registration. Article 12 has become a compromise in so far as it allows the establishment of a certain level of equivalence between a national (or international) standard and EMAS. Any additional demands imposed by EMAS on top of the other standard must be added later, and it is the verifier's task to check for the missing elements.

In fact, the Commission has been faced with a proliferation of standards: the British standard BS 7750 (published in April 1992; final version in March 1994), an Irish standard (IS 310), a Spanish standard (UNE 77–8012(2)-93 in 1993) and a French draft standard (NF X 30–200, published in spring 1993; valid on a testing basis until April 1995). In 1996 the Commission recognised those parts of the Irish, British and Spanish standards which correspond to EMAS (EC 1996a), and these have been used ever since (although, as is shown later in Table 10.4, only the British standard is widely used). In response to the prevailing uncertainty created by this proliferation, the Commission recognised the necessity of having only one harmonised standard and gave CEN a mandate to develop a European standard (EN) in 1994. At the same time,[3] members of CEN objected that European standards would cause them difficulties because of the global links of European companies. They saw a danger that these companies would have to apply different standards when operating abroad. They declared that it would be more convenient for them to participate in the development of an international standard, and asked to be allowed to get involved in the work of the ISO, which had initiated discussion of standards on environmental management systems and audits.[4]

Most member states were in favour of an ISO standard and saw advantages in having environmental management systems harmonised worldwide. Other member states, namely Denmark and Germany, were worried that the ISO standard might become too weak to be used as a European standard for EMAS purposes. At the end of the process, the international standard was to be adopted also as a European standard. The Vienna Agreement envisages a parallel voting procedure which allows for the simultaneous adoption of the standard text under ISO and CEN, provided that the texts do not differ in their specifications. The Commission neither agreed to nor refused the development of an international standard but pointed out that the mandate for a strictly European standard had to be met.

Members of CEN continued to join the ISO meetings, but their participation proved to be extremely difficult because they had to achieve as close a correspondence as possible between EMAS requirements and ISO standards, the binding specification document on environmental management systems[5] and the non-binding auditing guidelines.[6] They could not and did not want to risk rejection of the standard as insufficient by the Article 19 Committee and the Commission. Mainly, however, they negotiated for a standard as close as possible to EMAS requirements because, on one hand, they feared that comparatively strict EMAS obligations would place their industry at a competitive disadvantage to foreign firms and, on the other hand, they were threatened by non-EU standardisers that any CEN standard going beyond ISO would be regarded as a trade barrier (for more on the General Agreement on Tariffs and Trade (GATT)/World Trade Organisation (WTO) dimension, see Chapters 1 and 9; Vogel 1998; Golub 1998).

The concepts of ISO 14001 and EMAS differed from the outset: The ISO standard wants to achieve improvement in environmental management systems, whereas the goal of EMAS is the continuous improvement of environmental performance. This has not changed now that the standard has been adopted: 'it is important to understand and acknowledge that the ISO 14001 environmental management system is a systems approach to managing environmental issues and not a performance-based document' (Dodds 1997: 7).

As discussed in more detail below, EMAS is more demanding than ISO 14001 because it requires, among other things, a validated public statement, detailed data concerning the issues to be covered in the environmental policy, programme and audit, well-defined audit frequency and the inclusion of past activities.

Well-defined, strict standards, however, were not in the interest of representatives from third countries, especially the US. On the contrary, these states followed the diametrically opposite aim: they wanted to keep the standard as vague as possible because they were afraid that clearer terminology could be used against them in litigation, on the basis of the rather strict civil-liability legislation found in the US. Though private standards are of a voluntary nature, not applying them if one has subscribed to them can be used as an assumption of tort against the defendant. In fact, this anxiety appears somewhat misplaced, as US companies have to fulfil much stricter demands in order to receive insurance cover. They had to change their attitudes profoundly when civil liability for environmental damage was introduced, long before ISO discussions on ecoaudits began. According to one American expert on environmental liability:

1 US insurance companies denied insurance cover for past sins and current practices presented uninsurable exposure for pollution damages.
2 Environmental management systems cannot take credit for reopening the insurance market in the USA – at least not the type of environmental management systems that we are talking about when we refer to ISO 14001 or EMAS or BS 7750. Although many factors were involved, the use of legal compliance audits, site management audits, risk assessment techniques, and due diligence investigations (at the time of mergers and acquisitions) gave insurers confidence that they could write policies for enterprises. . . . Companies established full-blown environmental management systems and procedures to help ensure that they remained in compliance with rules, regulations and laws. These systems tend to look like ISO 14001, i.e. policies, procedures, training, documentation, and so on. . . . US companies have to do much more (than EU companies) because of the liability provisions. . . . One insurer scoffed at ISO 14001 in the sense that it adds nothing for those who underwrite risks to the routine practices already in place at US companies.[7]

Europeans, however, have no legislation on civil liability for environmental damages. EMAS is the first attempt to introduce measures for effective risk control on sites. The position of the Europeans became stronger with every new round of negotiation, especially following the intervention of the US Environmental Protection Agency (EPA), which insisted on introducing EMAS-like elements into the binding ISO 14001 standard, for instance 'commitment to legal compliance' (EMAS: 'provision for legal compliance') and 'commitment to prevention of pollution' (EMAS: 'continuous improvement of environmental performance').

The strong economic and legal concerns of the respective parties are understandable; compared to EMAS, ISO 14001 was in fact less demanding since its wording was less clear. In assessing the environmental merits of ISO 14001, a study by the US consultancy firm Benchmark, commissioned by the European Environmental Bureau (EEB), raised five provocative questions (EEB 1995):

1 Benchmark criticises ISO 14001's lack of commitment to Agenda 21 or any international environmental convention, and how it restricts itself to laws and regulations applicable in the respective country.
2 ISO 14001 can become an international trade standard without operative participation from governments or NGOs because, although observers from governments and NGOs may participate in the negotiations, only national standards institutes have voting rights.
3 A company certified according to ISO 14001 cannot demonstrate that it has good environmental, health and safety performance because environmental performance as conceived by this standard relates only to the

measurable performances of the environmental management systems. ISO 14001 is a specification standard for verifying only conformity with an organisation's own environmental policy, not of environmental performance in general.

4. In particular, Benchmark criticises the lack of public access to information. The only provision for transparency is that companies 'shall consider processes for external communication'.
5. ISO 14001 does not require transnational corporations to meet the environmental standards of their home country everywhere in the world. Rather, companies only have to apply local/national standards – a very disappointing result for an international standard.

The process of developing ISO 14001 was accompanied by the activities of an ad hoc working group from CEN's side which was composed of representatives from national standardisation institutes, governments, industry and an observer for environmental NGOs (the author). This working group compared the two texts – EMAS and ISO 14001 – in order to spot inconsistencies and to prepare the European delegation for the subsequent ISO meeting.

This was also an opportunity for the NGOs to learn about the working mechanisms of standardisation bodies. The standardisation institutes take over de facto legislative tasks, although they are not democratically legitimised and are not neutral because they are dominated by industry. This is often perfectly acceptable, but not when adopting standards for a policy like environmental management systems. Also, national standardisation institutes often delegate their representation to persons coming from industry, so it could happen that a Canadian from Exxon acted as a Belgian representative. Finally, although ISO should work according to the consensus principle (i.e. unanimous agreement after long discussions) the ad-hoc working group started voting, a procedure which by its very nature neglects minority positions.

The description of the standardisation process is crucial because there was a risk that ISO 14001 could replace EMAS requirements via Article 12 and so undermine a promising environmental instrument. Environmental organisations were concerned that ISO standards might serve as a pretext to amend the EMAS Regulation in order to prevent trade barriers, with the positive EMAS elements lost in the process.

Differences between EMAS and ISO 14001

An important difference between EMAS and ISO 14001 is that for EMAS the five Benchmark questions can be answered in a more positive sense. For instance, under EMAS international conventions must be followed as far as the EU has ratified them and transposed them into Community legislation.[8]

Also, unlike the ISO standard, EMAS was adopted in a more democratic decision-making procedure through the Council of Ministers after consulting the European Parliament. The most important difference is the aim: instead of merely validating a firm's adherence to its own policy goals (system performance), EMAS measures environmental performance and requires continuous improvement. Oversight and transparency are further improved through the obligation EMAS imposes on registered sites to issue a public statement. Sites have to meet not only local and national but also EU standards, since EMAS is only valid inside the EU.

Industry representatives used to contend that the EMAS requirements were implicitly contained in ISO 14001. This would certainly be true if a company was ambitious in respect to the environment. There is, however, no guarantee that companies will always be of good will, undertaking unilateral and ambitious investments towards environmental improvement. The dilemma lies in the language, which often leaves too much scope for avoiding important obligations:

- Where ISO 14001 remains vague in its terminology, EMAS is clear.
- Unlike ISO 14001, EMAS audit-checks for improvement of environmental performance rather than environmental system performance.
- The EMAS system is based on the results of the initial review.
- ISO 14001 speaks about 'environmental aspects' and not about 'environmental effects' or 'impacts' as EMAS does.
- ISO 14001 envisages a 'commitment to legal compliance'. A 'provision to legal compliance' (EMAS) is stronger, i.e. a firm will need to indicate timescale, as well as human and financial resources. 'Provision to legal compliance' is more than what many sites currently achieve. This requirement will make them act responsibly and gives them enough flexibility to plan for their compliance.
- ISO has laid down no requirement concerning audit frequency and the relevance of past activities.

The commitment to 'continuous improvement of environmental performance' is promising because sites, once they have entered the scheme, will have to demonstrate their dedication to the targets of EMAS, which become ever more demanding over time. Like an upward escalator, there is only one way to go if they do not want to lose their registration.

These main differences must be bridged if an industrial site certified under ISO 14001 wants to be registered under EMAS. Article 12 provides for a site certified to a standard to be given credit for its achievements under that standard. The missing elements, however, have to be accounted for. The CEN ad hoc group on EMAS thus made an inventory of the loose ends in the so-called 'Bridging Document'. The intention was to add this document as a non-binding CEN report to the CEN standard on environmental management systems.

The fact that this would have added additional requirements to ISO 14001 (in essence creating a new standard) raised many concerns in industry, which fears distortion of global competition: either Europeans could claim they had done more than other users of the same standard and were entitled to more credit, for example in a case of public procurement conditions,[9] or they would incur additional costs from meeting the demands of something like 'ISO plus' without receiving any benefits. European industry wanted, under any circumstances, to avoid giving the impression of deviating from the international standard. Therefore industry did not want to give the Bridging Document any binding status.

When ISO 14001 was finally voted on in the parallel procedure in 1996, CEN members voted in favour of the ISO standard, but in the parallel vote Germany and Ireland opposed adopting it as the CEN standard to serve EMAS under Article 12 of the Regulation. They found the differences between EMAS and ISO 14001 were too important. This did not bode well for the vote on the Article 19 Committee.

In light of these developments the EU Commission changed its strategy. The Bridging Document was dropped, but the Commission decided that the missing or less evident elements of ISO 14001 should be taken into account by the certification bodies and, later, the verifiers. It had always been the intention that the certification body would have to use the Bridging Document when delivering evidence that the site met its requirements, one of which was adding these missing elements for EMAS purposes. The verifier will assess the presented evidence before signing up.

Issues of certification

Certification bodies have to be accredited by the national (private) accreditation body, which follows certain rules to make sure that certification bodies are competent, independent and objective.[10] The EN 45012 standard profiles quality-certification bodies and the work they are accredited to deliver.[11] No standard yet exists for environmental certification bodies. Since the Commission was afraid it would take too long to write a standard to describe its brief, the federation of national accreditation bodies (EAC) was asked in 1994 to develop a guidance document which could be used to accredit environmental certification bodies. EN 45012 served as a basis and was adapted to environmental needs; the term 'quality system' was replaced with 'environmental management system'. The EAC guidance is then appended under each clause of the standard.

Representatives from national accreditation bodies negotiated the text, and, again, government, industry and NGO representatives participated as observers. The environmentalist who followed the negotiations regards the system as reliable and credible. The product of the joint deliberations was *EAC Guidelines for the Accreditation of Certification Bodies for Environmental Management Systems* (GAC); the final version was published in 1996. This text is not a standard. It is, in principle, an agreement among national private bodies which commit themselves to placing identical demands on the private certification bodies they will be accrediting. It is by no means a binding instrument and will be regulated by market forces. Given this situation, the German government has insisted on developing an international standard for the accreditation of certification bodies, but this will require several years of work.

The GAC defines many aspects of certification bodies, including their object and field of application, their general requirements, organisational structure and working procedures. A certification body must have a governing board which is responsible for performance, and which formulates and implements the policy. The senior executive is responsible to the governing board. It is very important that certification bodies are competent, impartial and non-discriminatory. The GAC describes independence as follows: 'the senior executive . . . has to be free from control by those who have a direct commercial interest in the products or services concerned' (clause 4b, p.9). Documented procedures enable the audit, certification and surveillance of environmental management systems, and – last but not least – withdrawal and cancellation of certificates.

The national accreditation bodies, the EAC members, are in charge of accrediting certification bodies. National accreditation bodies all have their own procedures and fee structures. Common procedures demanded by the EAC guidelines are assessment of the applicant's head office by document review and, most important, 'witness auditing' by the accreditation body on site. The scope of the certification body – the industrial activities they want to specialise in – will be defined. Accredited certification bodies will undergo routine surveillance by the accreditation body on both head office and site activity. National accreditation bodies will likely enter into a multilateral recognition agreement to recognise each other's work. Such an agreement will involve peer review. For the time being, the International Accreditation of Certification (IAC) is negotiating, on the basis of ISO/IEC (International Electrotechnical Commission) Guide 62,[12] a corresponding Guide for EMS, trying to transfer as much wording as possible from the European Guide on EMS.

The description of the accreditation process and the tasks of the certification bodies raises the question of whether these private institutions will be able to provide reliable results. The argument is that this corresponds to the system of chartered accountants for financial audits which our whole economy relies upon. Accreditation and certification bodies which do not act responsibly would fall out of the system because they have lost their credibility. Whether this applies for environmental certification bodies still remains to be seen. Analogous experiences of undeserved certificates obtained for the quality and safety of products give reasons for concern.

The EAC Guide demands that the certifier should not have provided consulting services to the site he or she is certifying. This does not preclude consulting firms offering both services to the same site as long as they guarantee that the activities are executed by different people. One might doubt, however, whether the two tasks can ever be entirely unrelated.

There is another concern: certifiers accredited to certify management systems for product quality are applying to accredit environmental management systems without really being qualified. EMSs differ considerably in their requirements from quality management, since the former is directed at the environmental impact of production methods, whereas the latter is directed at controlling product quality by setting up adequate management structures. Certifiers may have received additional training but they have not always been shown to satisfy the requirements of certifying EMS.[13]

In light of all these things, one should expect initial certifications to differ considerably in quality. Hopefully, over time, peer review, competition and a vigilant public will help to improve a system which has much to offer all stakeholders. None the less, the question remains: since the accreditation bodies will accredit certification bodies for the use of ISO 14001, how will differences from EMAS be handled?

In 1997 the Commission presented two texts to the Article 19 Committee on the recognition of the international standard ISO 14001 and ISO EN 14001. ISO 14001 was recognised to meet certain parts but not all provisions of the EMAS regulation, as were certification procedures which follow EAC guidelines, German and Austrian laws. As mentioned above, where ISO and ISO EN require an environmental *management system* audit the Regulation demands an environmental *performance* audit.[14] Through these two Commission texts, the system audit has found its way into EMAS.

Issues of verification

Unlike the certifier's brief, the requirements for verifiers are part of the Regulation. Verifiers have to prove their competence in assessing environmental performance in the industrial sectors they have special experience and knowledge of, as they are accredited for special NACE sectors of industry only. Furthermore, they must have the competence to conduct audits. They must be independent, objective and impartial. They will be accredited and supervised by a national system for which member states have 'to guarantee . . . independence and neutrality' and 'ensure appropriate consultation of parties involved' (Article 6(1); EC 1993a). 'Environmental verifiers accredited in one Member State may perform verification activities in any other Member State' (Article 6(7); EC 1993a). As shown in Table 10.2, the numbers of accredited verifiers (which totalled 131 in early 1997) varies considerably among the different member states, with Germany by far the frontrunner. Member states had to set up two competent bodies, one for the accreditation of environmental verifiers and the other for the registration of EMAS-validated sites. All member states – with the exception of Italy – have notified the Commission of their competent bodies. The accreditation of environmental verifiers lies in the hands of national accreditation bodies.

A closer look at the German system reveals that it is different because the German law implementing the EU Regulation includes an accreditation system for verifiers/certifiers (BG 1995: 1591). This law defines the terms of reference for the accreditation body for verifiers. As the German law envisages no separate procedure for certifiers, the verifiers will do both jobs. In Germany and France single persons are allowed to be accredited verifiers. The French and the German systems are criticised because individuals are recognised as verifiers, whereas in other countries they come as a team with a variety of competences. Doubts have been arising whether individuals can cover such a broad range of competences. The verifier in Germany, however, will contract accredited experts who possess special knowledge in a sector of industry. Austria has also implemented parts of the Regulation through a number of ordinances which make no distinction between verifiers and certifiers. The Article 19 Committee has decided that the different procedures are in compliance with the requirements of the Regulation.

The Commission has the task, together with the Article 19 Committee, of promoting collaboration among member states in order to avoid inconsistencies in the accreditation and supervision of environmental verifiers acting in other member states than their own. Furthermore, the Commission has provided a document to give environmental verifiers guidance when accom-

Table 10.2 Number of accredited EMAS verifiers in EU member states

	As of 13 September 1996	*As of 29 January 1997*
Austria	8	12
Germany	103	116
Denmark	3	3
France	6	9
Finland	2	2
The Netherlands	3	3
Spain	0	1
Sweden	3	5
The UK	7	7

plishing their task (EC 1995). The document is not binding, only of an informative nature. It will be continuously completed over time. It contains detailed interpretation of the text of the Regulation in order to assist verifiers on their job. This guide also describes independence. Unfortunately, a provision which originally allowed verifiers to validate a site only three times in succession was later dropped. A verifier must make sure that his or her work does not depend on one single company.

Some member states have set up very sophisticated procedures to accredit verifiers: in Denmark they are examined 'on the job', i.e. while they do an audit, under the surveillance of officials of the Environmental Protection Agency. In Germany, NGOs have participated to describe the requirements and qualifications of environmental verifiers.

It will be up to the verifier to make sure that the elements missing from other standards which are needed to fulfil the requirements of the EMAS Regulation have been provided by the certification body. Industry has made it quite clear that the verifier should not reopen the assessment of what the certification body has already provided. This puts the verifier in an awkward position: the verifier has to sign up with his or her name, but if he or she reopens the audit the verifier 'will be out of business', as one representative of a standards institute put it. Hopefully verifiers will be able to protect their own independence and credibility in the first place because that is primarily what they make their living on. The problem may be somewhat hypothetical, however, since very often certification and verification will be carried out by the same person.

The verifier is a key figure in EMAS because he or she is the mediator between the inside and the outside, the site and the stakeholders. Nobody else will be as exposed to the public as the verifier: while he or she has to check the performance of the site, the verifier also acts on behalf of the client, who has chosen and is paying him, and thus expects a certain loyalty, but when he or she signs the environmental statement he is accountable to the public.

The Commission regularly distributes handouts on EMAS registrations, information from which is summarised in Table 10.3. The number of registered sites is increasing rapidly, and about 70 per cent of them are German.

Since ISO 14001 is not yet fully operational many sites use BS 7750 as their standard. Industrial sites in most countries enter EMAS via a standard, whereas sites in Austria and Germany base themselves on the text of the Regulation directly. The difference between Germany and the other member states concerning the number of registered sites has caused much speculation. As a matter of fact, as is shown in Table 10.4, many sites have certified EMSs under one standard or another but have not gone further to seek EMAS registration: in the UK 156 sites had been certified under the British Standard 7750 and only 15 had an EMAS registration by November 1996, whereas in Germany 45 had been certified and 260 were registered under

Table 10.3 Certified EMAS sites in Europe

	As of 21 March 96	*As of 13 September 96*	*As of 7 February 97*
Austria	6	25	46
Belgium	2	2	2
Denmark	3	4	13
Germany	113	290	348
Finland	0	3	4
France	3	5	7
Ireland	1	1	2
The Netherlands	3	7	11
Sweden	1	5	43
The UK	9	20	25
EU 15	141	362	501

Table 10.4 Certified and EMAS-registered sites in the EU (as of November 1996)

	EMAS	*National standard BS 7750*	*ISO 14001*
Belgium	3	0	5
Denmark	3	25	25
The UK	15	156	156
Finland	2	0	6
France	3	0	5
Germany	260	0	45
The Netherlands	5	100	0
Ireland	1	8 (IS 310)	0
Italy	(pilot projects)	0	7
Austria	8	0	0
Portugal	(pilot projects)	0	0
Sweden	1	0	0
Spain	(pilot projects)	0	0

EMAS (UKOB 1996: 17). EMAS is also applied differently in spirit: 'The "soft" managerial style practised in the UK stresses cost effectiveness as the means for achieving environmental performance improvements. Germany pushes a strong line of environmental laws' (EG 1996: 7). For the time being no industrial site has been denied validation for not fulfilling the EMAS requirements. The first validation, however, has not yet assessed improvements; this will occur only when the next verification is due.

The environmental statement

A Swedish study which evaluated the environmental statements of 58 sites revealed a number of important problems (Swedish EMAS Council 1996). For a start, the length of the statements varied dramatically among sites, as shown in Table 10.5. According to the study, significant environmental effects, targets and objectives were clearly listed in 40 statements and not clearly listed in 15. Moreover, the types of pollution addressed in the statements also varied considerably (Table 10.6). Finally, there were enormous discrepancies in the method of verification: 2 statements were verified by a

Table 10.5 Length of environmental statements

Number of pages	1–10	11–20	21–40	>40
Number of sites	11	26	16	2

Table 10.6 Proportion of environmental statements summarising specific environmental issues

Issue	*Proportion (%)*
Pollutant emissions	76
Waste generation	100
Consumption of	
raw material	71
energy	100
water	95
Noise	38
Other environmental issues	45

validation stamp, 38 by a validation statement, 3 included full assessment and validation statements, and only 12 made reference to the verifier. Given these variations in length, content and procedure, one can only conclude that environmental statements did not live up to the demands of the Regulation.

The environmental statements are the only part of the EMAS process which is accessible. The statements should give information which allows the interested public to assess to what extent the site has improved its environmental performance. The first validation – ending with registration – considers only the existence of the EMAS elements, but progress should be measurable after the second validation at the latest. It would be valuable to have an indicator for progress for EMAS because otherwise progress might only be modest. The Regulation does not say anything about the speed at which companies must improve.

Evaluation of EMAS

EMAS could have much to offer if the framework is devised properly and registration of industrial sites achieves all the necessary requirements. It could help environmental policy move to a certain extent from end-of-pipe solutions towards a more integrated approach which focuses on environmental aspects of the production process as well as the products. Most of the pioneering firms which originally introduced management and audit systems proudly boast that, although they faced high initial outlays for the audit itself, they reaped substantial economic gains in the end, as the audit brought to their attention many important details of their production about which they had been previously unaware.

An Austrian study concludes that the introduction of EMSs is paying off (Austrian Economic Chamber 1996). Firms undergoing EMAS registration had earned their investment back after less than fourteen months, on average. The time required for implementing an EMAS system in a company was one person-day per employee. Some 60 per cent of the measures implemented in the course of the EMAS project were of an organisational nature; the remaining 40 per cent were related to technological matters. When evaluating the consequences of the measures applied, the firms rated the economic effects of the measures somewhat higher than their ecological impact.

First signals from Germany indicate that both banks and insurance companies are susceptible to the effort a company has made to obtain registration. In March 1996 the Deutsche Bank announced favourable rates of interest for EMAS-registered sites because it regards EMAS validation as a clear signal of reduced environmental risks. In addition, a number of important German insurance companies view the existence of EMAS registration as a favourable factor when assessing clients for their premium level. They insist, however, that they do not grant a 'rebate' because of EMAS (OB 1996b).

To take two more examples, the *Land* of Bavaria has negotiated an agreement with the federations of Bavarian industry and trade (UB 1995). They guarantee 500 EMAS-registered sites by the year 1999. The Bavarian government has declared its intention of working towards alleviating reporting and documenting requirements, controls and monitoring of regulatory authorities, as well as permit procedures for EMAS-registered sites. An Italian representative has stated that his country intends to implement the recently adopted EU Directive on Integrated Pollution Prevention and Control by applying EMAS.

Legal compliance

Public authorities could also have an interest in EMAS because there are limits to their ability to control all industrial sites. Some member states cannot afford the necessary qualified personnel and see scope, in theory, in controls carried out by private certifiers/verifiers, the costs of which are borne by the sites themselves. EMAS could be regarded advantageously for them, especially when the shrinking state budgets are considered, which will not even guarantee the maintenance of the – often insufficient – existing controls. As public authorities certainly do not plan to retire from controlling the sites, EMAS could help to maintain trust for sites which have convincingly demonstrated compliance efforts with previous legal requirements (Lübbe-Wolff 1996: 227).

As has already been indicated, local authorities are not yet very willing to give enterprises credit for their efforts. Examples have arisen, for example, where regulatory authorities have been especially severe with sites applying for registration and have been overly rigid in controlling them. The Commission is concerned that this attitude might discourage enterprises from registering. On the other hand, registered companies have also approached their regulatory authority with a request to have their licences reformulated in a more generous way, although this is obviously not in agreement with the requirement of continuous improvement of environmental performance.

To help governments interpret the Regulation, the Commission distributed a document to government officials which states:

> It is therefore clear that the EMAS Regulation does not introduce any idea of substitution between itself and environmental legislation, whatever the source is. Being voluntary, EMAS could not have this purpose. . . . This does not prevent the authorities in the member states, responsible for applying environmental legislation, developing a way to add value to participation in EMAS. This

> recognition of EMAS participation – usually called deregulation – is facilitated by the fact that registration in EMAS involves 'provision for compliance with all relevant regulatory requirements regarding the environment'.
>
> (EC 1996b: 1)

In the same document, the Commission also states under the heading 'EMAS and the implementation of environmental liability' that:

> The system does not generate any extra incentives for the enforcement authority to start pursuing companies. . . . In any event, it is the responsibility of the site to produce accurate information, irrespective of their participation in EMAS. The environmental statement does not provide a special opportunity to pursue a case of environmental liability [for a different view see chs 1, 5 and 6]. . . . On the contrary, the holistic approach of EMAS is a means to reduce the risks, thanks, in particular, to the better control of environmental effects it generates. This leads to changes from corrective actions to a preventive way of dealing with environmental issues.
>
> (EC 1996b: 2)

Industrial sites undergoing EMAS registration are not exempt from requirements under existing legislation. They only present evidence that there are grounds for the assumption that they have worked on their environmental impacts in order to reduce them. This effort by the industrial site should be understood as a demonstration of goodwill with the objective of creating a climate of trust. The only conceivable concessions which might be acceptable are where EMAS reporting requirements duplicate existing provisions.

This view is obviously not shared entirely by representatives of the International Standards Organisation with respect to enterprises which have been certified under ISO 14001. They answered the question 'What if an ISO 14000-certified company was found to be in non-compliance with an environmental regulation?':

> There would be a less severe reaction on the part of the regulatory authorities if the company were implementing an environmental management system.
>
> (Mr Renswik, Norwegian Ministry of Environment; quoted in Dodds 1997: 4)

> Things do go wrong and fall out of compliance. However, if something went wrong, a company implementing an ISO 14000-based environmental management system had procedures to deal with the occurrence. In addition, ISO 14000 was very clear in requiring the

company to declare its intent to comply with regulations, to do so and to be able to demonstrate that it is doing so.

(Mr Dodds, chairman of TC 207 Subcommittee 1, which developed ISO 14001; quoted in Dodds 1997: 4)

EMAS, subsidiarity and deregulation

Meanwhile, deregulation is high on the agenda in Brussels. The question arises whether EMAS is an instrument which can guarantee 'a high level of environmental protection', as demanded in the EC Treaty, at a time when the Community is reviewing its legislation and increasingly restricting itself to strategies and framework Directives to be filled in by member states following the principle of subsidiarity (see Golub 1996). Environmental organisations are worried by this because they foresee that the 'repatriation' of EU environmental legislation might entail eco-dumping. They fear eco-dumping, not in the sense that production moves away to 'pollution havens', but in the sense that bad examples may be imitated because industries do not want to follow stricter rules than their competitors in other member states and will lobby their governments to be entitled to produce according to the same (low) environmental standards. This problem was recognised twenty years ago and was one of the reasons that the Community created its environmental legislation.

This cannot be remedied by a system like EMAS, which has merits of its own but which are irrespective of the legal environmental norms. It will, by definition, not be the goal of EMAS to improve industrial sites beyond legal compliance. When the legal framework places inadequate demands on firms concerning emissions, waste generation, nature protection, etc., EMAS cannot be expected to repair these shortcomings. On the contrary, the stringency of data collection is not guaranteed in EMAS. There is a danger that the assessment might be too superficial and unsatisfactory, especially because the certifiers/verifiers might not be independent enough since the companies are their clients. It will be difficult to resist the temptation to please them in order to be invited to undertake the subsequent audit (Führ 1993).

Improvement through revision

The Regulation is weak where it leaves too much flexibility. This may have been a good principle at the beginning, but the text should become more explicit as time passes. When the Regulation is revised in 1998 the 'provision for legal compliance' should be changed into 'legal compliance', with provisions for temporary exemptions. The wording of 'commitment to continuous improvement of environmental performance, with a view to reducing environmental impacts to levels not exceeding those corresponding to economically viable application of best available technology' (Article 3a) leaves much to the imagination and substitutes good intentions for solid environmental improvements. The number of qualifiers reflects the difficulty of reaching a compromise during negotiations in Council. It should read instead: 'Registered sites shall improve environmental performance continuously and apply best available technology.'

Environmental organisations think that Article 12 is the weakest point in EMAS and would like to have it removed in a review. They publicise the advantages of EMAS over ISO 14001 and encourage third countries, including their environmental organisations, to demand some of the more stringent EMAS requirements when the international standard is reviewed in 2001. The ISO standard leaves even more room for manoeuvre, as it can be applied in a very demanding way but also rather superficially because of its vague wording.

EMAS registration does not stipulate many types of information in the public statement. It is up to the company whether it wants to disclose more information than is required under the Regulation. A revised Regulation should make the statement more demanding so that it will accustom industrial sites to regular reporting on the environmental impact of their activities. Furthermore, the information should be presented in a standardised form in order to make it comparable. The best way to issue emission data would be to relate them to production and not to give them in absolute amounts, as is currently done.

Civil liability for environmental damage

One of the most promising environmental instruments of all in Community legislation would be the introduction of civil liability for environmental damage, perhaps the only effective means of respecting the polluter-pays principle. Producers tend to internalise profits and externalise the cost to the environment in the form of pollution, waste, resource consumption and destruction of ecosystems. Civil liability would privatise the damage and send the costs of the environmental damage home to the one who has caused it.

The Fifth Environmental Action Programme envisages the introduction of civil liability for environmental damages. This project meets much scepticism concerning the insurability of the risks given the bad experiences in the United States, where insurance companies refused to insure against environmental damage once the law on environmental liability was introduced. The EU would be much better off in a comparable situation and EMAS will be of the utmost importance in this respect: companies which have set up environmental management and audit systems are more in control of their production processes than those without such systems, since they are better aware of their risks and can minimise or even avoid them, and so reduce their insurance premium. Thus, this is one of the greatest potentials of

EMAS: it could help to pave the way for industry to enter smoothly into civil liability for environmental damages. The advantage of legislation on civil liability is not that firms would run into more litigation (but see Chapters 1, 5 and 6). On the contrary, this is not the desired effect. They should take preventive measures in order to avoid damage, and EMAS is a first step towards organising internal management in a responsible way, although the efforts made under EMAS are still on the modest side.

Notes

1 Article 3(a) of the Regulation reads: 'must include commitments aimed at the reasonable continuous improvement of environmental performance, with a view to reducing environmental impacts to levels not exceeding those corresponding to *economically viable application of best available technology*.'

2 The CEN is the European Standardisation Committee, of which national standard institutes of the EU and the European Free Trade Association (EFTA) are members. The EAC is a federation of national accreditation bodies in charge of accrediting and supervising certification bodies.

3 The author was observer to the CEN delegation which discussed the terms of reference of the mandate with the Commission officials.

4 ISO had set up a new Technical Committee TC 207 (Environmental Management) in June 1993. In one of the six working groups of the committee, standards for environmental management systems and environmental auditing were discussed.

5 *ISO 14001 Environmental Management Systems – Specification with Guidance for Use.*

6 ISO Guidelines for Environmental Auditing:

- 14010 general principles for environmental auditing;
- 14011 on audit procedures;
- 14012 on qualification criteria for environmental auditors.

7 Personal letter from William D'Allessandro, executive editor, Victor House News Co. International, Amherst, NH.

8 The EU has ratified the Basel and the Biodiversity Conventions, while the US has not.

9 Access to public markets is one of the strongest motives for companies to undergo certification/verification under either a standard or EMAS. It had been one of the reasons to create the international standard from the outset, because companies overseas saw themselves disadvantaged compared to EMAS-registered firms and declared EMAS registration as a competitive advantage for EU sites which was in conflict with the GATT Agreement because it creates technical barriers to trade.

10 According to ISO definitions, certification is the 'procedure by which a third party gives written assurance that a product, process or service conforms to specific requirements' (ISO 1986a: clause 2), while accreditation is the 'procedure by which an authoritative body gives formal recognition that a body or person is competent to carry out specific tasks' (ISO 1986b: clause 2).

11 EN 45012 contains the 'General Criteria for Certification Bodies Operating Quality System Certification'.

12 This Guide contains the 'General Requirements for Bodies Operating Assessment and Certification of Quality Management Systems'.
13 Roger Brockway, UK Accreditation Board, personal communication.
14 Regulation No 1836/93, Annex II, E (1) includes the provision 'with the objective of evaluating environmental performance'.

References

Austrian Economic Chamber (1996) *Ecomanagement Pays Off: Evaluation of a Subsidised Programme for the Introduction of Environmental Management and Audit Systems (EMAS) in Austria*, commissioned by the Federal Ministry of Science, Transport and Arts, and the Federal Ministry of Environment, Youth and Family (Vienna: State Government of Upper Austria, Austrian Economic Chamber), June.

BG (1995) 'Gesetz zur Ausführung der Verordnung (EWG) Nr. 1836/93 vom 29. Juni über die freiwillige Beteiligung gewerblicher Unternehmen an einem Gemeinschaftssystem für das Umweltmanagement und die Umweltbetriebsprüfung (Umweltauditgesetz-UAG)', *Bundesgesetzblatt* 1(61), Bonn, 14 December.

Dodds, O. (1997) 'Environment: Getting to Grips with ISO 14000', *ISO Bulletin*, Geneva, January.

EC (1990) *Council Regulation (EEC) No. 3037/90 Establishing the Classification of Economic Activities in the European Community*, OJL 293, 24.10.1990 (Brussels: Office for Official Publications of the European Communities).

—— (1991) *Draft Proposal for a Council Regulation Concerning the Voluntary Participation by Commercial Enterprises in a Community Ecomanagement and Audit System*, COM(91)459 final; OJC 120, 30.4.1991 (Brussels: Office for Official Publications of the European Communities).

—— (1992) *Council Deliberations of the Environmental Council of 18 December 1992, Document of 18 December 1992 on the Draft Proposal COM(91)459 final*, on file with author.

—— (1993a) *Council Regulation (EEC) 1836/93 of 29 June 1993 Allowing Voluntary Participation by Companies in the Industrial Sector in a Community Ecomanagement and Audit Scheme* (OJL 168/1) (Brussels: Office for Official Publications of the European Communities).

—— (1993b) *Amended Draft for a Council Regulation Allowing Voluntary Participation of Commercial Enterprises in a Community Ecoaudit and Management Scheme*, COM(93)97 final (Brussels: Office for Official Publications of the European Communities).

—— (1995) *General Guidance Document for Accredited Environmental Verifiers Under Regulation 1836/93 and on the Verification and Validation Approach*, Doc. No. DG XI/821/95 (Brussels: Commission of the European Communities).

—— (1996a) *Commission Decisions 96/149–151/EC*, OJL 34, 13.2.1996 (Brussels: Office for Official Publications of the European Communities).

—— (1996b) *EMAS and Environmental Liability, 1. EMAS and Environmental Legislation*, Doc. No. DG XI/276/96 – VD, 4 July (Brussels: Commission of the European Communities).

EEB (1995) *ISO 14001: An Uncommon Perspective* (Brussels: European Environmental Bureau).

EG (1996) *Environmental Management Systems in Germany* (Sag Harbour, NY: Environment Group, Inc.), on file with author.

Führ, M. (1993) 'Betriebsorganisation als Element proaktiven Umweltschutzes', *Jahrbuch des Umwelt- und Technikrechts* 21: 145–87.

Golub, J. (1996) 'Sovereignty and subsidiarity in EU environmental policy', *Political Studies* 44(4): 686–703.

—— (ed.) (1998) *Global Competition and EU Environmental Policy* (London: Routledge).

ISO (1986a) 'General terms and definitions concerning standardisation and related activities', *ISO/IEC Guide* 2.

—— (1986b) 'Guidelines for third-party audir and registration of a supplier qulaity system', *ISO/IEC Guide* 48.

Lübbe-Wolff, G. (1996) 'Das Umweltauditgesetz', *Natur und Recht* 18(5): 217–27.

OB (1996a) *Ökologische Briefe 24*, 12 June.

—— (1996b) *Ökologische Briefe 16*, 17 April.

Swedish EMAS Council (1996) *EMAS Environmental Statement Survey* (Stockholm: Svenska Miljöstyrningsradet AB), April.

TEKES (1994) *Environmental Management and Auditing in the European Union, English Summary*, TEKES, Technology Development Centre Finland, distributed by the Embassy of Finland, Brussels, 30 May.

UB (1995) *Umweltpakt Bayern*, October.

UKOB (1996) *Umwelt, Kommunale Ökologische Briefe*, 27 November.

Vogel, D. (1998) 'EU environmental policy and the GATT/WTO', in J. Golub (ed.) *Global Competition and EU Environmental Policy* (London: Routledge).

11

ENVIRONMENTAL TAXES AND CHARGES IN THE EU

Jos Delbeke and Hans Bergman

Introduction

This chapter focuses on environmental taxes and charges within the EU. This issue is particularly interesting as it concerns not only a new dimension in environmental policy but also the domain of taxation, which remains the (almost) exclusive competence of the 15 member states. The next section deals with the relevant parts of the Treaty on European Union and the restructuring of environmental policy in the 1990s. It focuses on the institutional and policy framework in which economic instruments, as part of EU environmental policy, are being developed. Some specific examples in the fields of energy taxation and environmental levies are considered, as well as how the environment could be integrated into the broader context of comprehensive fiscal reform. The subsequent section considers the division of responsibility between the EU and its respective member states, and briefly describes the increasing use of new instruments at the national level. Then there follows a section which outlines some of the problems inherent in a proliferation of national instruments and discusses several mechanisms available for preventing distortion of the internal market, including the emerging EU legal framework surrounding the use of fiscal instruments in member states.

Redesigning environmental instruments at the EU level

When environmental legislation and harmonisation of the laws of individual member states started in the European Community in the 1970s it was mainly related to the functioning of the internal market. The Single European Act, adopted in 1987, gave EU environment policy a much firmer basis through a full chapter in the treaty (Articles 130R-T). That treaty introduced three fundamental principles which are of relevance when considering implementation of economic instruments such as environmental charges and taxes:

- prevention is better than cure;
- damage should be rectified at its source;
- the polluter should pay.

The next change to the legal framework governing environmental action came in the Treaty on European Union. Article 2 of the treaty, signed in Maastricht in November 1993, states: 'The Community shall have as its task . . . to promote . . . sustainable and non-inflationary growth respecting the environment'. The Fifth Environmental Action Programme (EAP), adopted by the EU in March 1992, set the framework for action, stressing as key objectives the integration of the environment into economic policies and the broadening of the range of instruments.

The institutional discussions on the future of the European Union and the continued economic recession of the mid-1990s have led to a fundamental reflection on the nature of the Union's environmental legislation. Part of this discussion relates to the question of whether and how the cost-effectiveness of environmental policy measures could be improved. Most European environmental policy has been produced fairly recently and has in many cases been prompted by ad hoc pressures. The result is a body of some 200 environmental laws which are based mainly on technical regulation. This represents the starting point for consolidation and improvement. Recent criticism, evidenced clearly by the Molitor Report, has focused on two main elements: the cost-effectiveness and coherence of the legislation in place (European Commission 1995; Collier 1997). In other words, the fundamental message is more focused on reregulation than deregulation. It is generally recognised that deregulation would not be in line with the consistent message the European citizen has been giving, for example through Eurobarometer polls, in which environmental policy and employment creation continue to appear as the two most important objectives for the EU.

A key theme in this reflection concerns the need for a sharper distinction between objectives and instruments. In particular, the use of environmental quality objectives as explicit policy aims is likely to increase in the coming years. It is becoming essential to define more precisely what is meant by 'clean' air, 'clean' water, etc. At the European level, a framework Directive (96/62/EC) on ambient air quality was adopted in 1996. Daughter Directives on sulphur dioxide (SO_2), nitrogen oxides (NO_x, lead, particulate matter and ozone are being prepared, each of which incorporates extensive reliance on economic evaluation techniques.

An equally important aspect of restructuring EU environmental policy is the commitment to a broader range of instruments (European Commission 1992, 1993, 1994, 1996). The attractions of the use of economic and, particularly, fiscal incentives are multiple (see Chapter 1). In principle, they reduce the costs of compliance for industry, and therefore the consumer.

They provide flexibility for industry in their response to environmental objectives and a continuous incentive for technological innovation. They also provide revenues which can be used to reduce other more distorting taxes. Taxes might be levied on input material, energy, products, emissions and waste. Depreciation allowances can be varied so that investments in clean technology are depreciated over a shorter period of time than is normal.

There has been wide recognition and acceptance of the idea that tax reforms are required, shifting tax burdens away from labour and capital towards natural resource consumption in order to achieve the so called double dividend – simultaneous environmental and economic improvement (see Golub 1998). The idea was developed in the 1993 White Paper of the Commission on *Growth, Competitiveness and Employment* (COM(93)700). As is discussed below (and in the other chapters of this volume), significant tax shifts of this kind have already occurred in some EU countries, mostly centred around energy and carbon taxes.

The relevance of energy taxation for Community policy concerns its ability to promote the use of more efficient technologies in general, and to create incentives for consumers and producers to use cleaner fuels, not least because the more polluting fuels are usually cheaper to produce. The picture is, however, complex from a European Union point of view: fifteen member states, with different structures of primary energy use and endowment of energy resources, with different industrial structures and degrees of economic development, and, finally, with different environmental priorities and problems. In this respect, the European Commission has defined climate change and acidification as environmental problems where common action may bring added value to the policies member states are implementing, because of their transboundary or global nature. Many local environmental problems, such as urban air quality, are caused by products which are extensively traded within the internal market (e.g. cars and fuels); hence they are also more efficiently tackled through common initiatives at European level.

The carbon/energy tax proposal

The carbon/energy tax proposal was adopted in 1992 as part of a comprehensive strategy presented by the European Union at the United Nations Conference on Environment and Development in Rio de Janeiro (Heller 1998; Skjaerseth 1994). The joint Energy/Environment Council decision of October 1990 to stabilise carbon-dioxide (CO_2) emissions in 2000 at 1990 levels was integrated in the Climate Change Convention. Recent simulations and forecasts confirm that the European Union is unlikely to reach its stabilisation target in the year 2000 without the use of an appropriate fiscal instrument.

Tax rates were proposed equivalent to $US3 per barrel of oil equivalent at the outset, with annual increases taking them to $US10 after seven years. Several specific provisions were incorporated with a view to mitigating the possible impact on industrial competitiveness: among other things, special treatment for energy-intensive industries was provided for, and the revenues raised (+/-1 per cent of EU GDP) were to be used to offset other taxes.

The intensive discussion in the institutions of the European Union centred on two themes, the competitiveness impacts on industry and the increased fiscal competence of the Community. On the former theme, energy-intensive industry opposed the measure, fearing loss of market share. Any decision would, according to industry, have to be followed up by analogous evolutions in the world, or at least within the Organisation for Economic Cooperation and Development (OECD) region. However, the decisive factor preventing the adoption of the carbon/energy-tax proposal has been the unwillingness of some member states to increase the tax competence of the European Community.

Transport

The regulatory model to control transport-related air pollution followed by the Community has been based primarily on setting emission standards associated with vehicle-emissions control technology (e.g. catalytic converters) and with fuel quality (e.g. lead in petrol). This is, fundamentally, a normative and technical model which establishes emission-limit values that must apply across the European Union, and depends on the existence of effective programmes of inspection and maintenance of vehicles to control their compliance with the standards.

With a view to improving and updating existing legislation, the so-called auto-oil programme was created in 1993 as a cooperation venture between the motor-vehicle industry, the oil industry and the Commission. The objective was to generate sufficiently reliable data and technical analysis to be used in the preparation of two proposals for Directives, one on motor-vehicle emission standards and one on the quality of petrol and diesel fuels. The proposals were adopted by the Commission on 18 June 1996 (COM(96)248) and are still under discussion in the European Parliament and the Council.

The auto-oil programme incorporates the question of how to ensure a cost-effective set of new proposals. To date, the regulatory approach has not made as much use of non-technical measures and economic incentives as their cost-effectiveness would imply.

A potentially cost-effective, non-technical policy measure is based on tax differentiation. Recent experience in Europe shows the effectiveness of tax differentials to facilitate the introduction of vehicles equipped with new emission control technology or the introduction of cleaner fuels. In the EU the percentage of unleaded petrol sales, for example, increased from below 1 per cent in 1986 to, on average, 66 per cent in 1995.

The role of fiscal incentives, differentiated or not, is not limited to facilitating the adoption of new vehicle technologies or cleaner fuels. Damage to public health, ecosystems and buildings, for example, caused by motor-vehicle air pollution is primarily a social cost, not covered by road users. Economic efficiency suggests that this social cost should be internalised, which is equivalent to the implementation of the polluter-pays principle. The Green Paper *Towards Fair and Efficient Pricing in Transport* (COM(95)691) adopted by the Commission suggests the need to reform vehicle and fuel taxation to better reflect their respective environmental costs.

Comprehensive reform

Finally, tax 'rationalisation' – the need to ensure that environmental considerations are included in tax rules introduced in the past – is becoming an increasingly important issue. Cases in point are the regulation that kerosene for aviation fuel should be exempted from taxes, and the environmental problems related to the expansion of air transport. As a matter of fact, about 15 per cent of the external costs of air pollution in Europe can be attributed to air transport. As in the case of cars, the question is whether environmental taxation is able to reduce air emissions more cost-effectively than standards or technical devices.

A consultancy study conducted on behalf of the Commission, which examined environmental implications of tax systems, identified significant scope for action (European Commission 1997). One conclusion of the study is that the current value-added tax (VAT) system offers room for improvement regarding the rational consumption of water and energy. The study also concludes that urban sprawl and the use of company cars are being favoured through privileged tax provisions. Changes in some of these provisions could have positive effects for the environment and favour efficiency in resource allocation.

Thus, a comprehensive strategy to enhance the cost-effectiveness of environmental policy must examine the consistency of current tax provisions vis-à-vis environmental objectives. In this field one cannot expect things to change overnight. However, there would be progress if policy-makers in the member states considered the environmental implications of the exemptions and derogations in their tax systems.

Once an environmental objective has been defined, either at EU or at member-state level, the problem of determining the most cost-effective method of reaching it remains. The Commission supports an approach whereby, in most cases, the choice of instruments is determined at national level. This is often rational because ecosystems and geographical conditions vary amongst member states. Moreover, preferences for instruments may differ. An objective may be reached in one place through technical regulation, while elsewhere levies or negotiated agreements may be preferred. In the past, however, a common thread has been the emphasis placed on the definition of EU-wide technical standards, mostly known as best available technology (BAT). The use of technical harmonisation was seen as an essential means of avoiding unnecessary competitive pressures within the internal market (see Chapter 1).

Increasingly, however, it has been stressed that instruments other than technical harmonisation may be more cost-effective in solving environmental problems. Therefore much stress has been put on broadening the range of instruments available at the national level, mostly as a complement to the technical regulation. The following section briefly examines such developments in several states.

New Instruments in member states

Since the early 1990s there has been an increase in the use of environmental levies and charges in the member states, for example on fertilisers, pesticides, packaging, batteries. This increase has in many ways led to a substantial improvement in attaining environmental objectives.

At national level, there are interesting initiatives, along the lines proposed in the 1993 White Paper on *Growth, Competitiveness and Employment*, advocating a reduction of indirect labour costs (about 1–2 per cent of GDP), to be financed by other taxes like carbon and energy taxation, or energy taxes in general.

Belgium	Energy tax on households to finance employers' contributions to social-security schemes. A Ministerial Commission is preparing a variety of ecotaxes on products (see Chapter 5).
Denmark	General tax reform 1994–1998 decreased labour taxes to 2.7 per cent of GDP and increased ecotaxes to 1.4 per cent of GDP. Energy package 1996–2000 increases taxes on industrial energy use by 0.2 per cent of GDP, partly recycled as reduced non-wage labour costs.
Netherlands	Energy tax on small-scale users mainly recycled as lower labour taxes. The Green Tax Commission is looking at a possible 'greening' of the tax system (see Chapter 4).
United Kingdom	Landfill tax recycled as lower social-security contributions (see Chapter 2).
Luxembourg	Cuts in employers' social-security contributions financed by increased energy taxes.
Sweden	The Green Tax Commission is looking at a possible 'greening' of the tax system. Increases in CO_2 tax rates.

Such national initiatives are evidence of the interest that member states have in developing their economic and fiscal policies in complementary directions, thus diminishing the distortions of the fiscal system at the lowest cost and enhancing the effectiveness of environmental policy. In this respect, it should also be noted that member states are under pressure to stabilise or reduce their budget deficits with a view to further integration, particularly monetary union.

National diversity and preserving the single market

Given the continuing need to maintain a level economic playing field, member states considering implementation of environmental levies (such as taxes, charges, etc.) have sometimes been faced with an apparent contradiction between the environmental objectives of the treaty and the other treaty objectives – most notably the functioning of the internal market and technical harmonisation.

It is useful to distinguish between environmental levies on products and emissions. While levies on emissions in principle only affect domestic firms, for example factories situated in that member state, levies on products can affect both domestic and foreign producers. Hence, levies on products tend to be more sensitive from an internal-market point of view. The potential problems between levies and trade arise from the possibility of discrimination against imported products compared to domestically produced goods, in an open or hidden way, and through a number of means:

- A higher levy on imported than domestically produced products. Such discrimination in an open fashion is prohibited under both EU and international trade law.
- A levy on products which are largely imported while close substitutes produced domestically do not bear the levy. Such a case could be defensible, for example if the domestic products have a similar function but a better environmental effect. However, such a system could also be misused for purely protectionist purposes.
- In the packaging field it is increasingly common for levies to give preference to reusable packaging compared to one-way (recyclable) packaging. Such levies have created controversy, as it is often easier for a producer close to the local market to organise a deposit/return or bring-back system.

Levies often cause resistance, and have a tendency to give rise to complaints to the Commission. Often, both the complainants and the defenders (member states) refer to EU legislation to support their views. After discussion with involved parties – member states, complainants, etc. – some modification of tax laws often leads to a situation that satisfies all parties. There have therefore not been any fundamental European Court of Justice (ECJ) cases on environmental taxes or charges.

Traditional legal constraints on national experimentation

When member states wish to use environmental taxes there are a number of different EU laws they need to be aware of, which are embodied in the treaty and in secondary legislation such as Directives and Regulations. The main relevant principles of the treaty are expressed in Articles 9/12, 30/36, 76, 92–93, 95, 99, 100a, 130R-S. In summary, these articles aim to ensure that competition within the single market is not unduly disrupted by:

- customs duties or charges having equivalent effect;
- quantitative restrictions on imports or exports of goods between member states;
- state aid constituting distortions of competition affecting intra-Community trade;
- internal taxation discriminating against imported products or otherwise protecting national production.

Protection of the environment is a legitimate objective of general interest and one of the main objectives of the Community as a whole as spelled out in Article 130R.

Quantitative restrictions

The borderline between customs duties and taxes is of relevance because the treaty contains an absolute prohibition on customs duties and charges with a similar effect to customs duties, while it only contains a prohibition against taxes which are discriminatory. If the revenues from a tax are used to offset fully the burden for domestic producers, the charge will be assessed as a customs duty or charge with similar effect. If the tax falls only on imported products it can be assessed as a tax if the charge system is part of a general internal unbiased taxation system.

As regards the borderline between taxation and quantitative restrictions, environmental taxes in general fall within the scope of Article 95 (taxation), which normally excludes the application of Article 30 (quantitative restrictions). However, there are two situations where Article 30 would apply. First, Article 30 would apply in the absence of any similar or competing domestic production and in so far as the levy is of such an amount that the free movement of goods is impeded. Second, if the measures consist of several conditions or factors which are not necessarily linked to the levy itself or its proper functioning, such factors or conditions – for example labelling requirements – may then be assessed under Article 30.

If Article 30 is applicable, the protection of the environment is recognised as a so-called 'mandatory requirement', which may justify the measure even if it would hinder the free movement of goods (ECJ case on Danish bottles, Case C-302/86, 20/9–1988). In such a case the following conditions shall be met:

- The measure must be non-discriminatory.
- It should be shown to be necessary in order to meet the environmental objective.
- The burden which the measure imposes should be proportionate to the objective of protecting the environment.

Internal taxation

Article 95 aims at guaranteeing the complete neutrality of internal taxation as regards competition between domestic and imported products. However, differentiated taxation of imported and similar domestic products is legal provided the taxation is non-protective. Indeed, the ECJ has ruled (14 January 1981, taxation of denatured alcohol, Case 46–80, [1981] ECR 77) that Community law did not restrict the freedom of each member state to lay down tax arrangements which differentiate between certain products on the basis of objective criteria, such as the nature of the raw materials used or the production processes employed.

Proportionality, that is, balancing the gain for the environment with the potential impact on the single market, is not applicable when assessing an ecotax according to Article 95 except, as stated above, when it concerns administrative control measures of the levy.

The definition of a 'similar' product is important in determining whether a tax is discriminatory. The case law takes into account not only the objective characteristics of products but also whether they satisfy the same consumer needs. The consumer impression of a product is thus an important aspect when assessing compatibility with Community law. By adopting a Regulation on ecolabelling the Community has implicitly recognised that ecologically adapted products are not similar to products with the same function but with different ecological properties. The ecological difference could be embodied in the product or could be due to differences in production methods (this regulation is discussed in Chapter 9).

The case law of the court also indicates that the mere fact that a tax is levied predominantly on imports is not enough to deem the tax discriminatory. The court ruled that a tax which imposes heavier charges on a certain product than on another one on the basis of the raw materials used and the manufacturing process employed is not a violation of Article 95 if it is applied identically to the two categories of products.

Even when an environmental tax is based on objective and non-discriminatory criteria it may be necessary to take into account how the revenue from the charge is used in order to make a complete assessment of its effect on the internal market. When the revenue from a charge is used to partly offset the burden borne by domestic products the charge constitutes discriminatory taxation within the meaning of Article 95 of the treaty (Case C 17/91, [1992] ECR I 6523).

State aid

One area of concern to many member states is the distributional impact of environmental levies. One way of addressing these concerns and increasing the acceptability of environmental levies is by using the revenues for specific purposes (see Chapter 1). For example, revenues can be redistributed to those who paid them, but in accordance with a set of different criteria than those defining the payment. Alternatively, the revenue may be used for specific environmental purposes such as support for environmental investments or financing the collection of dangerous waste. However, the use of revenue can also have negative effects. According to the treaty, state aid to firms is, in principle, not allowed. Permission must be sought from the Commission. Handing back revenues from environmental levies, either as investment support or in other ways is considered to be state aid, and thus must fulfil certain requirements.

Under Article 92, any aid granted by a member state which distorts or threatens to distort competition by favouring certain undertakings or the production of certain goods shall, in so far as it affects trade between member states, be incompatible with the common market. Revenues from levies constitute 'state resources', and their use may therefore constitute state aid in the meaning of Article 92. Exemptions from an environmental tax might also constitute aid. The principles according to which aid schemes pursuing environmental objectives shall be assessed by the Commission are set out in the Community guidelines on state aid for environmental protection (OJC 72, 10.3.1994, p. 3), as will be further discussed below.

Secondary legislation on indirect taxation

Community legislation adopted under Article 99 contains:

- harmonised rules on tax structure and minimum rates for excise duties on mineral oils, tobacco and alcoholic beverages;
- other general provisions in Directive 92/12/EEC allowing member states to introduce indirect taxes on products provided that those taxes do not give rise to border-crossing formalities in trade between member states.

An important provision of Directive 92/81/EEC (OJL 316, 31.10.1992: 12) is that, in general, only one tax rate per product can be used. Directive 92/12/EEC (OJL 76, 23.03.1992: 1) provides member states with the possibility, within certain restrictions, to introduce other national taxes on mineral oils. Such taxes must comply with the rules applicable for excise duty and VAT. Member states may request authorisation from the Council to apply reduced tax rates or exemptions, for example for environmental reasons. Granted derogations concern tax differentiation such as reduced rates on environmentally improved diesel, and on reformulated unleaded and leaded petrol. In addition, as discussed below, member states have the right to apply reduced tax rates or tax exemptions on mineral oils used for specific purposes, for example in the field of public transport and within the agricultural sector.

Notification

The above examples show how Community law affects the use of environmental levies within the internal market. The levies can potentially harm the functioning of the internal market, for example by being misused for protectionist purposes, without creating benefit for the environment. At the same time it is clear that these levies can be very efficient in pursuing environmental policy. Thus it is important for the Commission to strike a balance between the environment, trade aspects and avoiding misuse of environmental levies. The issues involved are often complex, resulting in a number of cases involving intense discussions between representatives of the environment and industry.

An important information tool is the notification requirement. This means that member states in general have to inform the Commission before introducing a new measure. Through this procedure the Commission can study the proposal to ascertain its compatibility with Community legislation. In the process other member states are also informed and have the opportunity to comment. For measures considered to be state aid, including exemptions from new taxes, notification is compulsory. A state-aid scheme put into operation without being notified is illegal, and the Commission may order the member state to suspend the aid programme to give time for investigation. Should the member state refuse, it risks being brought in front of the ECJ. So far this has not happened. For environmental taxation the notification requirement is more indirect. Directive (83/189/EEC) contains a requirement for member states to notify technical standards and regulations which are linked to fiscal or financial measures affecting the consumption of products. Normally this Directive requires a standstill period of three months, giving time for the Commission to investigate, but this requirement does not apply to the fiscally related technical regulations.

If the Commission finds that any element of a new law does not fulfil the requirement of the treaty it can demand that the member state postpone implementation until certain changes in the legislation are made. Should the member state refuse, it again risks being brought in front of the ECJ. During such standstill periods informal contacts between the Commission, the member state concerned, other member states which may have interest in the case and complainants take place. The objective is always to try to find a solution which satisfies all parties. So far no case has been taken to the ECJ by the Commission.

The emerging EU framework

While the treaty and secondary legislation provide a certain amount of guidance, a substantial level of uncertainty remains over how much scope national governments have to devise and apply new environmental instruments. A central objective of the Commission is therefore to provide member states (including European Economic Area (EEA) states) with clarification on controversial issues related to the use of environmental levies and to provide guidance on existing possibilities to use them, as regulated by EU law. This section discusses several recently adopted pieces of an emerging EU framework, and describes some of the ways in which more formal, written guidelines for new national instruments might alter the relationship between the Commission and the member states.

The recognition of the need for a variety of instruments to reflect differing situations leads to new questions which have so far gone unanswered. As discussed above, the treaty and its related legislation create some limitations on the freedom of action of the member states, in particular where the functioning of the internal market may be affected. Hence, there is a need to spell out the legislative framework with which new environmental policy instruments can be introduced at national level. Obvious candidates for such initiatives are guidelines covering environmental taxes, voluntary or negotiated agreements, and environmental liability rules. This section looks only at the first of these categories.

Subsidies

Explicit guidelines exist on environmental state aid provided by member states to firms (OJC 72, 10.3.1994). The guidelines were developed in 1993 in order to clarify for member states for which objectives and to which intensity aid is allowed. Before the guidelines existed there were often long discussions on whether different aid programmes were in conformity with state-aid regulations.

The guidelines basically state that state aid to support environmental investments is allowed, between 30 per cent and 40 per cent, only if the investments aim at achieving environmental protection at significantly higher levels than those required by mandatory standards. However, for plants older than two years, state aid of between 15 per cent and 25 per cent may also be authorised to comply with mandatory requirements. The guidelines also indicate that exemption from environmental taxes, which is regarded as state aid, may be allowed if it is necessary to prevent domestic firms being placed at disadvantage compared with their competitors in countries that do not have such measures.

Environmental agreements

In the *Communication on Environmental Agreements* (COM(96)561) of 27 November 1996, the Commission presented guidelines for the use of agreements between public authorities and industry in the field of the environment. The background for this initiative is the principle of shared responsibility and the need to broaden the range of policy instruments to achieve a better instrument mix.

Environmental agreements can be legally binding, with obligations for both parties, but they may also be non-binding 'gentlemen's agreements' in the form of a unilateral commitment recognised by the public authority. They can bring about effective measures in advance of legislation, and thus reduce the volume of regulatory and administrative actions. In addition, environmental agreements can encourage a proactive approach from industry and are likely to lead to cost-effective measures, because they allow industry to adjust environmental investment to their medium-term capital investments. They can also be more quickly implemented than regulations, which is an important advantage in areas with fast technological developments.

On the other hand, not all of the past 'voluntary' agreements were transparent and credible (see Chapter 8). The Commission therefore suggests that interested parties should be consulted before an agreement is concluded and that agreements should, wherever possible, be binding on the parties. They should go beyond unspecified 'best-effort' clauses and include quantified targets. Clear monitoring measures should be defined. Third-party verification is also suggested, as well as publication of the agreement and of the results achieved.

While agreements are instruments which can be used at local, regional, national, Community and international level, the competencies of the European Community only relate to agreements concluded at Community level and to national agreements used to implement Community Directives. Neither aspect is completely new to the environmental policy of the Community, but the preconditions for the use of agreements need to be further clarified. Based on the Communication, the Commission issued a Recommendation concerning how environmental agreements can be used in implementing Community Directives (OJL 333, 21.12.1996: 59).

Environmental taxes

In 1997, the Commission published a Communication entitled *Environmental Taxes and Charges in the Single Market* (COM(97)9). The document supports environmental taxes and charges in the member states that are used in a way compatible with Community legislation. The document therefore explains the legal framework applicable to member states, and clarifies both the possibilities and constraints for member states to act in this field. The document mainly deals with product taxation, as this is the area most sensitive to internal market aspects. It is explained that the effects of the European legislation, among other things, are that:

- if a levy has a clearly positive environmental effect it may be judged in a more positive way in terms of its effect on other policy areas;
- levies may not be used to discriminate against products from other member states;
- levies should be in accordance with secondary legislation on indirect taxation, for example in the field of energy taxation, where detailed rules exist;
- exemptions from paying the levy and the way revenues from environmental levies are used should fulfil rules in the field of state aid.

The document also specifies when member states have to inform the Commission of their activities (notification rules). Such rules exist in the following areas:

- state aid;
- technical standards and regulations linked to fiscal measures (Directive 83/189/EEC);
- national measures taken to transpose Community Directives into national law.

As this is a rapidly evolving area, the Commission will closely follow the evolution of the use of environmental taxes and charges in member states and their impact on the single market and on environmental policy. The Commission plans to carry out an evaluation on the economic and environmental effects of their use. The results of this work will be used to draw policy conclusions on the further use of environmental taxes and charges on Community and member state levels. However, this does not mean that the Commission will not make proposals in this field before the evaluation is finalised.

Energy and fuel taxes

A substantial amount of work has gone into developing EC guidelines in the area of national energy and fuel taxes. The carbon/energy-tax proposal, together with the globalisation of the economy and the increased competitive pressure on European industry, provoked a fundamental debate on the future of taxation in the European Union. This debate was launched by the European Commission in the previously mentioned White Paper on *Growth, Competitiveness and Employment*. It was shown that over recent decades a significant erosion of tax revenues has taken place, originally related only to capital income.

The trend of the last few years, to increase tax revenues from labour, is no longer considered an appropriate way to compensate for this loss. Such a policy increases indirect labour costs and hinders the creation of new jobs, hence contributing to the historically high and persistent unemployment figures in the European Union. In this context, the member states rely increasingly on two types of indirect taxes to counter the structural erosion of their tax receipts: value-added taxes to be paid on all final consumption and specific taxes on certain goods, for example energy products.

The member states of the European Union have a long history of energy taxation, in particular on mineral-oil products. This form of taxation was not originally motivated by environmental concerns. Governments tend to see energy as a stable fiscal base because of the relatively inelastic demand. Since the early 1980s energy taxation has increased slightly, from 2.1 per cent to 2.6 per cent of GDP for the EU as a whole. Researchers have concluded that the side-effects of this type of taxation have been beneficial for the environment: it has been shown that it contributed substantially to the higher overall fuel efficiency of the car fleet in Europe compared to the United States. Today, consumer taxes on motor fuels, i.e. excise and value added tax, constitute some 65 per cent – 75 per cent of the final price (see Table 11.1). Thus, member states have a powerful tool, not only to raise revenue, but also to differentiate according to environmental performance.

The situation described above represents an important evolution. First, discussions on the carbon/energy tax have made it clear that, in the current institutional context, member states prefer not to create a new harmonised tax system. Thus, for the time being the requirement for unanimity in the Council of Ministers when voting on fiscal issues is set to stay. Second, despite the existence of this institutional constraint, member states nevertheless need a Community framework to develop their fiscal policies, in particular on product taxation, as goods can be freely traded throughout the internal market without border controls. As a consequence of both considerations, the Commission was asked by the member states in 1996 to develop a comprehensive Community approach towards the taxation of energy products, based on experience with the system of excise duties.

Table 11.1 Petrol taxes in the EU

	Euro 95					*Diesel*				
	Price	*Tax*	*Final price*	*% tax*	*% ad valorem*	*Price*	*Tax*	*Final price*	*% tax*	*% ad valorem*
Belgium	220	632	852	0.74	2.87	231	413	644	0.64	1.79
Denmark	211	613	824	0.74	2.91	215	428	643	0.67	1.99
Germany	203	628	831	0.76	3.09	225	411	636	0.65	1.83
Greece	187	459	646	0.71	2.45	180	328	508	0.65	1.82
Spain	212	469	681	0.69	2.21	201	348	549	0.63	1.73
France	170	730	900	0.81	4.29	175	464	639	0.73	2.65
Ireland	226	459	685	0.67	2.03	268	411	679	0.61	1.53
Italy	222	634	856	0.74	2.86	217	474	691	0.69	2.18
Luxembourg	211	430	641	0.67	2.04	198	332	530	0.63	1.68
TheNetherlands	231	676	907	0.75	2.93	234	427	661	0.65	1.82
Austria	262	568	830	0.68	2.17	255	411	666	0.62	1.61
Portugal	225	557	782	0.71	2.48	200	341	541	0.63	1.71
Finland	219	717	936	0.77	3.27	225	405	630	0.64	1.80
Sweden	233	663	896	0.74	2.85	319	448	767	0.58	1.40
The UK	164	506	670	0.76	3.09	175	508	683	0.74	2.90
EU 15	213	583	796	0.73	2.73	221	410	631	0.65	1.85

On 12 March 1997 the Commission adopted a new energy-tax proposal (COM(97)30), to be submitted to the Council and the European Parliament for decision. This proposal for a Directive intends to modernise the Community system for the taxation of mineral oils and extends its scope to other energy products such as coal and gas. The Commission is thus meeting the obligation contained in Article 10 of Directive 92/82/EEC to review the minimum rates of excise duty on mineral oils. It is also responding to the Council's request, expressed following the deadlock in negotiations on the CO_2/energy tax, that it should present new proposals in the field of taxation of energy products. Finally, it takes into account the European Parliament opinion on the Commission's report on minimum excise-duty rates (COM(95)285, 13.9.95), which asks that it define a consistent basis for taxation covering both mineral oils and competing products.

In addition, the Commission has recently presented a draft Directive laying down technical specifications which fuels to be put on the market in the Community must satisfy (COM(96)248). This will trigger new investments and operating costs for the oil industry, estimated at Ecu13.2 billion over fifteen years. It would be unnecessarily costly to set up such standards at a level well beyond what most local environmental conditions within the EU might require. Thus the proposed Directive may require the marketing of higher-quality fuels.

Community framework legislation in this area was therefore necessary to allow member states to use tax-incentive schemes for cleaner fuels. It allows member states to provide fiscal incentives in favour of cleaner fuels. Such differentiation within a Community framework can enhance the market penetration of higher-quality fuels without disrupting the internal market. The internal market is undisrupted as the Directive specifies some cleaner fuels for which member states may provide fiscal incentives. In this way there may be two or three different fuel specifications in the EU, instead of perhaps thirty, which would be the case if each member state had two separate specifications.

Since its implementation, a number of member states have utilised the possibility offered by the Directive to apply tax differentiation in connection with objectives of environmental policy. In addition to the leaded/unleaded differential, six member states (Denmark, Greece, Ireland, Sweden, Finland) already apply or have requested permission to apply duty differentials based on the environmental qualities of fuels.

Car emissions

In the field of car emissions there are harmonised technical regulations in order to create an internal market for cars in which competition can be free and fair. The Community has defined a legally required technical standard which all new cars must fulfil. Before this standard was mandatory it was known for several years that it would become mandatory at a specified date. During this period a lower standard applied.

The Directive in question (94/12 EC) describes a precise framework for the use of fiscal incentives by the member states, via the sales tax, to favour cars fulfilling the stricter standards before they became compulsory (see Chapters 3 and 4). The Directive is thus an example of a harmonised framework, where each member state could work out the specific details of the environmental instrument according to its preferences and tax systems. At the same time, a well functioning internal market was maintained, as there was a maximum of two different technical specifications which car firms needed to adopt – the compulsory lower standard and, if they so wished, the higher standard to enjoy the tax discounts given by some member states.

The Community is currently working out future emission standards for cars. As this work is somewhat delayed, there is only one Community standard (which applies to all cars sold today). There is therefore no future standard that member states can give tax incentives to. Some member states currently give tax incentives for the purchase of new cars with technical specifications that go beyond those of existing Community law. This may not be fully in accordance with the Directive, but might be accepted due to the delay in the new Community legislation on future emission standards.

Conclusions

Pressures to reconsider the traditional command and control approach towards environmental policy have generated a lively debate in the European Union. This debate centres around three key issues: the requirements of the environment chapter of the Treaty on European Union, the safeguarding of sound competition within the internal market, and the recognition of the cultural, political and environmental differences among the member states.

Broadening the range of environmental policy instruments available at both the EU and national level remains firmly on the Commission's agenda. The deregulation trend in recent years reinforces, through its insistence on a cost-effective environmental policy, the critical role of a variety of instruments at the disposal of the member states. This evolution coincides with the desire of member states to follow their own preferences in the use of a particular instrument.

For reasons related to the different environmental conditions of member states and the different preferences of their population for environmental policy measures, one can, for the immediate future, expect most initiatives in the area of levies to be developed at the national level. According to the Eurobarometer 1995, national 'green' taxes receive a high degree of public support throughout the EU.

The existence of an internal market, however, has a tendency to impose a significant degree of convergence on the instruments chosen by member states. Furthermore, current Community legislation specifies some limitations to the use of particular environmental policy instruments. Thus the Commission's essential role is in developing explicit frameworks on the use of some instruments such as energy taxes and specifications on how to use taxes to encourage sales of environmentally improved cars and fuels.

The Commission has also issued more general guidelines, which show the room for national action on environmental policy instruments such as environmental taxes and charges, and on environmental agreements.

References

Collier, U. (ed.) (1997) *Deregulation in the European Union: Environmental Perspectives* (London: Routledge).

European Commission (1992) *Communication on Industrial Competitiveness and Environmental Protection*, SEC(92)1986, 24 November (Brussels: Commission of the European Communities).
—— (1993) *Towards Sustainability: Fifth Environmental Action Programme*, OJC 138, 17 May (Luxembourg: Commission of the European Communities).
—— (1994) *Communication on Economic Growth and the Environment: Some Implications for Economic Policymaking*, COM(94)465, 3 November (Brussels: Commission of the European Communities).
—— (1995) *Report of the Group of Independent Experts on Legislative and Administrative Simplification*, COM(95)288, 21 June (Brussels: Commission of the European Communities).
—— (1996) *Economic Incentives and Disincentives for Environmental Protection*, proceedings of the European Commission and Council Presidency Conference, Rome, 7 June.
—— (1997) *Tax Provisions with a Potential Impact on Environmental Protection*, study carried out by Moret and Ernst & Young for the European Commission (Luxembourg: Office for Official Publications of the European Communities).
Golub, J. (1998) 'Global competition and EU environmental policy: introduction and overview', in J. Golub (ed.) *Global Competition and EU Environmental Policy* (London: Routledge).
Heller, T. (1998) 'The path to EU climate change policy', in J. Golub (ed.) *Global Competition and EU Environmental Policy* (London: Routledge).
Skjaerseth, J. (1994) 'Climate policy of the EC: too hot to handle', *Journal of Common Market Studies* 32(1): 25–45.

INDEX